IoT Platforms, Use Cases, Privacy, and Business Models

Carna Zivkovic • Yajuan Guan • Christoph Grimm
Editors

IoT Platforms, Use Cases, Privacy, and Business Models

With Hands-on Examples Based on the VICINITY Platform

Editors
Carna Zivkovic
Department of Computer Science
TU Kaiserslautern
Kaiserslautern, Germany

Yajuan Guan
Department of Energy Technology
Aalborg University
Aalborg Øst, Denmark

Christoph Grimm
Department of Computer Science
TU Kaiserslautern
Kaiserslautern, Germany

ISBN 978-3-030-45318-3   ISBN 978-3-030-45316-9   (eBook)
https://doi.org/10.1007/978-3-030-45316-9

This Springer imprint is published by the registered company Springer Nature Switzerland AG
The registered company address is: Gewerbestrasse 11, 6330 Cham, Switzerland

Foreword

In providing a definition of the "Internet of Things," different experts will usually offer different answers. Indeed, the term IoT is rather inexact and has been adopted as a common term for many different technologies.

This book offers an easily understandable explanation of the fundamental terms, providing a solid knowledge base to "nonexpert" readers. On the other hand, it highlights several hot topics of the IoT ecosystems and explains them more deeply and should also provide interesting information for IoT professionals.

Thus, for example, the book explains the IoT-relevance of European privacy legislation GDPR, the applicability of homomorphic encryption to preserve users' privacy, the role of ontologies and IoT standards to achieve semantic interoperability of IoT ecosystems, the IoT-specific aspects of digital security, and many other specific aspects.

Since the authors of this book are from academic, commercial, industrial, and service-provider domains, they bring a broad range of expertise and practical knowledge in IoT relevant developments. Moreover, their experience has been deepened while they have been working together on the open-source IoT platform "VICINITY" for 4 years. Thanks to this, most of the IoT aspects that are addressed in this book are demonstrated using hands-on examples that are linked with concrete realistic use cases.

In addition to the theoretical knowledge, the book comes with "hands-on coding examples" that allow the reader to kick-start his/her own IoT applications, smart services, etc. based on the latest state-of-the-art technique assistance.

In summary, this book provides a coherent single-source introduction into the IoT and is recommended for students or beginners with only fragmented understanding and knowledge on IoT, as well as those who wish to build on their existing knowledge.

Chief Executive Officer   Stefan Vanya
BAVENIR, S.R.O. (BVR)
Bratislava, Slovakia

Preface

The Internet has been used by humans since decades to retrieve documents. The Internet of Things (IoT) extends it towards a network, where many different kinds of machines are networked with each other. Such a "Machine-to-Machine" communication permits us to establish new services that are able to change our world as they no longer require human interaction.

This book has been written to give the reader an overview and a practical introduction to the IoT. Mastering the IoT requires a domain-crossing understanding, from the networked, physical things up to the services and the driving business models. The book provides the reader with a comprehensive and consistent introduction to all relevant topics. The topics have been carefully chosen to allow the reader to get a holistic understanding of the IoT.

The selected topics include IoT platforms, use cases, business models, ontologies, IoT standards, the European privacy legislation GDPR, security and homomorphic encryption, and many other issues. The topics are carefully introduced and explained. For each topic, the reader gets a theoretical introduction and an overview. However, the book is not intended to be just a collection of theoretical knowledge. Where applicable, the theory is backed by brief coding examples. For this purpose, we use the IoT platform VICINITY that is open-source (https://github. com/vicinityh2020), free, and allows the reader to very quickly set up its own IoT devices and use cases.

Kaiserslautern, Rhineland-Palatinate, Germany   Carna Zivkovic
Aalborg Øst, Denmark   Yajuan Guan
Kaiserslautern, Rheinland-Pfalz, Germany   Christoph Grimm

Acknowledgments

The first big thank you, the authors would like to give, is to the VICINITY project partners. Without you and your hard work in the project, this book would have never become a reality. It was a pleasure working with you and we are sincerely looking forward to future opportunities for partnerships and collaborations.

We would also like to express our sincere gratitude to our project reviewers Mirko Presser and Aleksandra Bukala and project officer Joel Bacquet for the valuable feedback on the quality of our project work. We truly appreciate your time and effort to selflessly share your experience with us and provide us with valuable and helpful comments that have led to the significant improvement of the VICINITY IoT platform. We are more than happy that we met your expectations and we truly hope that we will have a chance to have you as reviewers in our future projects.

A special thanks goes to the European Commission for ranking the VICINITY project proposal as one of the best-written proposals and providing funding for its realization under the biggest research and innovation EU program, Horizon 2020.

Last but not least, we are extremely thankful to Springer, as our book publisher. A big thank you for doing an excellent job in reviewing and further shaping our book and giving it the opportunity to see the light of day.

Contents

Chapter 1
An Introduction to the Internet of Things

Johannes Kölsch, Carna Zivkovic, Yajuan Guan, and Christoph Grimm

1 From the World Wide Web to the Internet of Things

The internet has its roots in the 1980s. At that time, the more and more popular computer networks allowed users to access documents that were stored on computers across the world, for example, using the *ftp* protocol on the internet in universities. For private use, commercial but proprietary networks such as America Online (AOL) or the German Bildschirmtext (BTX) were popular. The proprietary nature and the heterogeneity of documents motivated the need for more standardised document formats and easier access in the late 1980s and early 1990s: Gopher [1] was an initially popular approach in that direction. Gopher integrated many features such as generated menus, formatted text, and references.

At around the same time, Tim Berners Lee worked towards a "World Wide Web" of documents. It was provided to the public *for free* in April 1993. It included the "Hypertext Markup Language" (HTML) as well as formatted documents, just like we are using it today [2].

Technically the term "internet" refers to the underlying lower-level communication protocols described in RFC 790-793 resp. 7323 (2014), and the term WWW

J. Kölsch (✉) · C. Zivkovic · C. Grimm
TU Kaiserslautern, Kaiserslautern, Germany
e-mail: koelsch@cs.uni-kl.de; zivkovic@cs.uni-kl.de; grimm@cs.uni-kl.de

Y. Guan
Department of Energy Technology, Aalborg University, Aalborg Øst, Denmark
e-mail: ygu@et.aau.dk

© Springer Nature Switzerland AG 2021
C. Zivkovic et al. (eds.), *IoT Platforms, Use Cases, Privacy, and Business Models*,
https://doi.org/10.1007/978-3-030-45316-9_1

refers to the HTML document format that can be considered an application layer protocol, hence the communication includes the following layers:

- The physical connection between two points is implemented by, e.g., Ethernet or Wireless Local Area Network (WLAN) (physical and media access layer).
- The term *internet* commonly refers to the transport of data in a network, using the protocols Internet Protocol version 4 (IPv4), Internet Protocol version 6 (IPv6), Transmission Control Protocol (TCP), and Internet Control Message Protocol (ICMP) (transport, network layers).
- The term *www* commonly refers to the application, that is, e.g., Hypertext Transfer Protocol (HTTP) resp. Hypertext Transfer Protocol Secure (HTTPS). In the WWW this is used to transport documents in the HTML format.

1.1   The Internet of Things

At the same time as the WWW was born, it became also clear that not only humans are using computers to access some documents: as well all kind of machines can use the internet to exchange data with each other, and create services. One of the first examples dates back to 1982, when a Coca Cola machine connected to a refrigerator via the Internet was able to report if there were cold drinks available [3]. This was by far not the only example. Mark Weiser described in 1991 [4] the vision of "ubiquitous computing", where computers are integrated into devices of our daily life and are just perceived by the services they provide—but not anymore as computers.

The nowadays popular term "Internet of Things", short IoT, is used today for the networking of smart objects via the internet so that these objects communicate with each other, and provide services without the need for inter-human or human-to-computer interaction. The term "Internet of Things" has been coined by Kevin Ashton [5] in 1999 in the context of Radio-frequency identification (RFID) tags. The International Telecommunication Union (ITU) defines the IoT as [6]:

a global infrastructure for the information society, enabling advanced services by interconnecting (physical and virtual) things based on existing and evolving interoperable information and communication technologies.

The number of devices connected to the internet and to each other continues to grow and according to the latest forecast from International Data Corporation will reach the value of 41.6 billion in 2025 [7]. As the connected devices in complex IoT networks are based on different technologies the first question that could be asked here is *How do we make them capable to interact with each other?* A number of IoT platforms available at the market today tries to answer this question. However, another question that arises here is how to decide what IoT platform to use and be sure that this is the right decision. The first step in the decision process is to look at an IoT reference architecture that allows us to map it to the capabilities of existing IoT solutions, and could also serve as a guideline for developing new products.

1.2   Communication in the IoT: Standards and Protocols

About 20 years ago the Internet slowly became mainstream. Due to the high adoption rate of personal computers combined with an Internet connection, computer science changed life because the information is available for everybody, at any time and (almost) without any latency, now. There are only a few basic principles of the Internet: data transport works per node to node connection, so routing is required to ensure reaching the desired destination. However, it takes only one connection to a node to use every service on the Internet. So every Internet-ready device can establish a connection with each other. In general, every node can be a service provider (server) or a consumer (client). Not forgetting one of the most important principles that every data sent through the Internet is handled equally. The above-mentioned principles are illustrated in Fig. 1.1, where the solid line depicts the chosen route to the destination server. The route follows the green marked dots. This implicates that the other dots and routes are not selected, but they are also a possible alternative route.

Anyway, to ensure a fluent use of the Internet various standard protocols such as TCP/IP and UDP (User Datagram Protocol) have been established. However, these standards cannot be directly applied in the IoT and M2M communications. Interconnected devices require protocols that support low power communication. In the following we provide a brief description of some of the widely used protocols in the M2M/IoT communication.

1.2.1   M2M/IoT Communication Protocols

MQTT (Message Queue Telemetry Port) MQTT is an open network protocol based on publish/subscribe principals. It is particularly designed to support the communication of devices with power constraints such as IoT devices. In the MQTT-based communication devices represent clients that are connected to an MQTT broker. The broker acts as a server that routes communication between clients. *How does publish/subscribe principle work?* A device that wants to distribute new message is called a publisher. The publisher sends data to a broker that distributes the message only to the clients that subscribed to the topic of the message. These

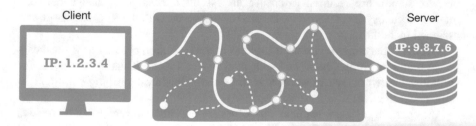

Fig. 1.1 Example of a connection between client and server including the route

clients are known as subscribers. As the protocol is based only on topics of messages distributed there is no need for a publisher to have any information about subscribers and vice versa. This makes the MQTT perfect for the application in low bandwidth and high latency networks such as IoT networks. This protocol is intended for centralised networks of IoT devices where a server controls all communications between devices. One common application of the MQTT protocol is the communication between an IoT gateway and devices connected to it.

CoAP (Constrained Application Protocol) CoAP is a specialised web transfer protocol designed for the use of resource-constrained devices on the Internet. This protocol is also known as a service layer protocol. It is defined by RFC 7252 and further extensions suitable for the M2M/IoT communications have been implemented by ETF CoRE (Constrained RESTful Environments) working group. In terms of transport, this protocol usually runs over UDP. Optionally, it can be bound to DLTS (Datagram Transport Layer Security) providing strong security capabilities for M2M communications. It is based on a RESTful model like HTTP: each resource is assigned a *Unique Resource Identifier (URI)* by the servers and clients use HTTP methods GET, PUT, POST, and DELETE to access these resources [8]. Like HTTP CoAP is not constrained in terms of data formats that can be used. The most widely used are XML (Extensible Markup Language) and JSON (JavaScript Object Notation).

1.2.2 TCP/IP

The combination of TCP (Transmission Control Protocol) and IP unlocks the potential of packet-oriented data transfer. It constitutes the backbone of the Internet because it fulfils fundamental communication requirements. In the 1980s the work of the Defense Advanced Research Projects Agency (DARPA) played a central role because they developed the standards of TCP/IP for their own prototype network called the ARPANET. The University of Southern California prepared RFC 760 for DARPA which specifies IP [9] and RFC 793 for TCP [10]. Following up Fig. 1.1 there are some derivations which can be made to identify more problems in Internet communication. First, data has to reach its destination travelling across lots of Internet nodes. Second, data has to be fragmented in single data packets of fixed size to guarantee transmission. Moreover, mechanisms to detect packet loss, duplicates, or permutation are required to ensure an essential message exchange between interconnected systems. These topics are fully covered with the RFC standards and assigned in Table 1.1.

TCP/IP connection represents a lossless communication channel as it reaches a server endpoint from a client endpoint over several nodes. Thus, it is suitable for centralised networks where a server is the main actor in establishing communications between clients.

The operating systems of client and server are providing the endpoints as sockets or the so-called socket-API (more in Sect. 1.2.6). Here a reliable stream is essential which is provided by TCP using special three-way handshake mechanisms [11].

Table 1.1 Differentiation of tasks of TCP and IP

Task	Standard
Logical addressing	IP [9]
Fragmentation	IP [9]
Routing	IP [9]
Error control	TCP [10]
Flow control	TCP [10]
Application support	TCP [10]

1.2.3 UDP

UDP which was defined by RFC 768 in 1980 is a communication protocol running at the Transport layer. It assumes that IP is employed as the underlying protocol. It sends messages which are referred to datagrams and provides a best-effort service which means an unreliable datagram protocol without delivery control, duplicate protection nor ordering functionality. Compared with TCP, preliminary communications/handshake mechanisms are not required to transfer data, thereby providing less-reliable but low-latency connections between applications.

A UDP datagram is kept in a single IP packet, therefore having maximum payload limits. To transmit a UDP datagram, the appropriate fields in the UDP header (PCI) are completed and the data together with the header for transmission are forwarded by the IP network layer [12].

1.2.4 Hyper Text Transfer Protocol

The HTTP is a generic and stateless protocol [11]. Hypertext stands for a text whose content links to another text document or in other words: a hypertext links with hyperlinks to other hypertexts. HTTP relies on a message exchange model of client and server, where the client sends requests and the server answers with responses like shown in Fig. 1.2.

Each request and response consists of a header and a payload. Headers include solely text, while the payload contains text or binary data. Referring to Fig. 1.2, the client requests data from the server and sends its request header with the desired parameters, codec, and path. Note that it could also be possible to directly send a payload with the request. The server answers the client with a specific response status (see Table 1.2) and the desired data, like an HTML document, as the payload. Through HTTPs stateless characteristics, the client has to provide all the required information necessary for the server.

According to the widely spread version HTTP 1.1 different methods are provided to make unambiguous assignments already in the request header. The most-commonly used are: GET, PUT, DELETE, and POST. In the case of calling the GET method a request should only retrieve data identified by the request-URI. The PUT method requests that a server updates a resource identified by the request-URI with

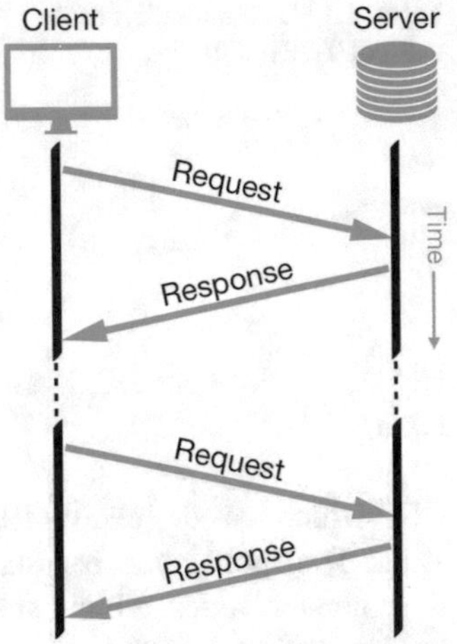

Fig. 1.2 Sequence diagram
of HTTP requests and
responses over the time

Table 1.2 An extract of
HTTP status codes of RFC
2616 [13]

Status number	Message
1xx	Informational response
100	Continue
2xx	Success
200	OK
201	Created
202	Accepted
3xx	Redirection
301	Moved permanently
4xx	Client errors
403	Forbidden
404	Not found

the data sent in the request body of the HTTP PUT request. The DELETE method
requests that the resource identified by the request-URI is deleted by the server. The
POST method requests that a server creates a resource identified by the request-URI
with the data sent in the request body of the HTTP POST request.

1.2.5 What Does Universal Resource Identifier (URI) Represent?

An URI is important to observe an abstract or physical resource. An URI is built up
out of a sequence of characters fragmented in components with special operators.
However, the generic URI syntax is hierarchical and each component has its
meaning to reach the resource. In general, an URI syntax consists of the following
components:

```
URI = scheme "://" host : port "/" path "?" query "#" fragment
```

Table 1.3 Decomposition of an URI

Component	Meaning
Host	IP address or a registered name (e.g., hostname) of a host located on a server
Port	Port number of the host
Path	A sequence of segments separated by a slash that identifies the resource in the host that the client wants to access
Query	An optional component preceded by ? mark providing information that can be used for some purpose (e.g., search)
Fragment	An optional component preceded by # mark, only locally processed by the web-browser that allows additional identification of the resource in the host

Meaning of each component is given below:
Example: http://10.0.0.1:80/news/?q=test#X_new

The blue marked components are used to communicate as mentioned in the TCP/IP section. The components marked in brown are part of the HTTP request (Table 1.3).

1.2.6   REST-API

API refers to M2M communication and it provides the common base for the client-server interaction. It is a fundamental task of distributed connected systems to communicate with each other while exchanging compatible data. Therefore various paradigms for remote procedure calls exist. Note that a REST-API is neither protocol nor standard, moreover it is an implementation style called RESTful architecture. Nevertheless, REST operates with the standardisation of URI, HTTP and usually uses JSON as a data format. Therefore a REST-API follows six principles. In contrast to other approaches of APIs, REST does not process any specific method in the URI. Instead, it just follows the URI to the resource and then processes the given HTTP request there. Due to the close resemblance of REST with HTTP, each visited URI with a static content is already conform with the REST architecture.

It is worth mentioning that a RESTful architecture is not intended for lots of content changes in a very short period of time as each change requires a new HTTP request [14].

1.2.7   JSON Data Format

In RFC 4627 JSON is described as *"a lightweight, text-based, language-independent data interchange format. [. . .]. JSON defines a small set of formatting rules for the portable representation of structured data"*.

So it provides a text format for the serialisation of structured data. Furthermore, JSON is a derivation of the JavaScript programming language. To start from a top-down point of view, a JSON file persists out of objects and arrays. A JSON text is also serialised as an object or an array. Objects are unordered collections of name or value pairs. And arrays are an ordered sequence of values pairs. A name pair is a string, while a value pair can be a string, number, Boolean, null, (nested) object, or an array [15].

2 IoT Reference Architecture

With the increased complexity of IoT, there is a need for structuring its architecture in layers. Zhu et al. [16] and Chen et al. [17] introduced three layers required to build an IoT reference architecture:

- Things layer
- Network layer
- Application layer

The main actors of each IoT network are smart things, sensors, and devices that implement the lowest level of the IoT reference architecture, *things layer*. This layer is also known as sensing or perception layer [16]. In the rest of the book, we will use only the term *things layer*. The things need to be connected and to communicate with each other via the internet. This is the responsibility of the network layer. Last, but not least the interaction with the end-user is needed. The application layer is the place where this interaction takes place.

Things Layer The main responsibility of this layer is to produce data that will be collected and transferred via the network layer. It uses different communication standard technologies such as RFID, ZigBee, Bluetooth, and 6LoWPAN. This layer is also responsible to transform the protocols of these technologies to a common communication protocol required for communication and connectivity at the higher layers; the protocol transformation is usually implemented in universal devices, the so-called IoT gateways. The gateways can be seen as interfaces between things layer and the second higher layer, i.e., the network layer.

Network Layer This layer is also known as communication or connectivity layer and it is responsible to transfer data collected from the things layer to remote destinations via Internet communication protocols such as Ethernet, WIFI, and GPRS.

Application Layer This layer lies on the top of IoT architecture and interacts directly with end-users. It delivers various application services to meet the users' needs; depending on the needs this layer covers different application domains such as smart home, smart transportation, smart energy, eHealth, and smart building.

Although these three layers are fundamental, they are not sufficient. The heterogeneity and the complexity of IoT systems grow continuously and it is hardly possible to manage all processes in IoT without additional layers. Further services are required to deal with a huge amount of data produced by devices supporting different technologies. And this is where a four-layer comes to place [18]. The layer is called *service management layer* and it is added on the top of the network layer.

Service Management Layer The main purpose of this layer is to enable service discovery and manage services and their requests by using appropriate identifiers like URI. It is also responsible to store data collected from the network layer, create semantic information out of it, perform information discovery using semantic technologies. It is also responsible for making smart decisions and further information processing up to the application layer.

Now, the questions we can ask here: Ok, when we created an IoT network how can we make a good business model out of it? To answer this question we need the fifth layer, the so-called business layer [18].

Business Layer As *service management layer* manages services, the responsibility of this layer is to manage applications, view information produced by the applications, and create business values out of it.

This leads us to a 5 layers IoT architecture [18, 19] that covers all crucial domains for building an IoT system: object domain, network (connectivity) domain, middleware domain (data storage, data analysis, service management, semantic services), application and finally, business domain, as shown in Fig. 1.3.

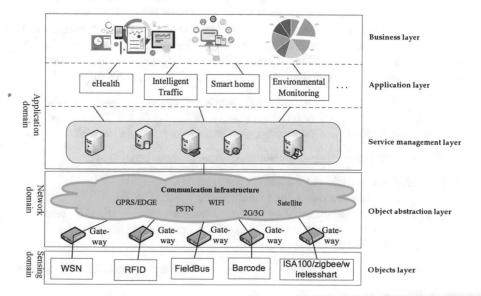

Fig. 1.3 5-layer IoT architecture [20]

3 Use Cases of the IoT: Beyond a Fridge with Webcam

In this section, we explain typical use cases of the IoT and the services that the IoT implements in these use cases.

3.1 IoT Services: How Much? and Where?

The IoT is often misunderstood as a network in which devices are networked. However, the IoT goes beyond that. Let us start with two simple examples from daily life to better understand which functions make IoT use case useful, and where:

(1a) *Mary is at work. She has forgotten to switch off the lights at home and should go back to switch off the lights. With an IoT light bulb, Mary can use an app on her mobile to switch it off.*

In example (1a), Mary uses the internet to control a device without her physical presence. Another example could be:

(2a) *Mary is shopping for a party. She does not remember which drinks she needs for the guests. Her fridge shows her the drinks inside, and she can get the missing ones.*

In example (2a), a device senses and sends data to Mary on her request. The above examples deal with low-level communication. But they lack important aspects: aggregation of data, reasoning, and automatic control. The IoT permits the collection, aggregation, and analysis of a vast set of data: sensors in various places, from personal data, and from social networks permit to give quite comprehensive and surprisingly good information on a situation. These permit devices not only to react to human interaction as in (1a) and (2a), but to take the decision, considering the pragmatics of a well-understood situation. Hence, the examples might be:

(1b) *Mary is leaving to work. The IoT concludes based on several sources of information that nobody is at home, and switches the light off. In the evening, her daughter asks her why there are so strange switches on the wall.*

In (1b), Mary has saved time and resources. Some of these benefits are:

1. Control of devices without the human presence (Ex. 1a),
2. Getting data without a delay due to human interaction (Ex. 2a),
3. Aggregating data and information to support decision-making (Ex. 1b), and
4. Automatic control of resource utilisation by a combination of information, as shown in the example below.

The more sources of information can be joined, the more pervasive will be the impact:

(2b) *Mary takes place in her Google car. It decides to bring her to a shop because she has a party on her calendar and she has not enough drinks in her fridge. It selects a shop nearby and shows her the drinks that her friends liked on Facebook.*

The above examples show the benefits and, clearly, the risks of IoT technology. We do not want to give up control of our daily life and our private data as in (Ex. 2). However, the example shows us the huge impact that the data aggregation can have on IoT services.

3.2 Use Cases of the IoT

Figure 1.4 shows a selection of domains in which the Internet of Things is used. The list is by far not comprehensive; we give a selection of the most popular ones.

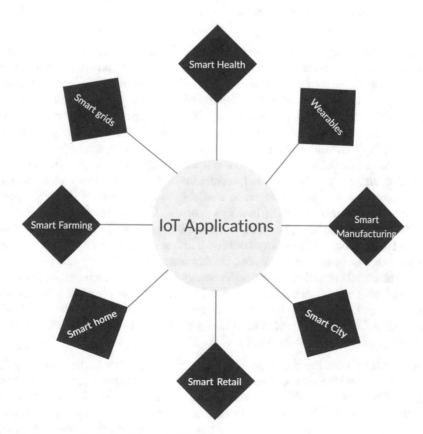

Fig. 1.4 Some applications of Internet of Things

Energy The meet of distributed energy generation systems with IoT gives rise to the Internet of Energy which is the vision of the smart grid. The integration covers not only the electrical grid infrastructure, but also all aspects of the grid including the electrical devices, sensors, actuators, appliances, their interactions, as well as high-level applications and use cases that are based on their data.

Interconnection of distributed energy systems within a neighbourhood can allow them to negotiate as a group their potential forecasted energy flexibility (both consumption and distributed generation) within a smart grid ecosystem, allowing the realisation of collective dynamic Demand Side Management (DSM) strategies. The use cases can target the management of a community-scale microgrid implemented within the municipal setting. The microgrid serves with a cluster of Municipal buildings with its power and heat requirements. The Energy flows through the microgrid-enabled energy network, from both the Generation and Demand sides, must be managed and maintained at minimum cost and maximum technical efficiency levels. Effective management can only be obtained from knowledgeable and informed decisions. This will be achieved by gathering data from sensors and energy system information models that will reduce the amount of time needed to elaborate on a decision.

Buildings and homes It is important to improve resource management, resource consumption, and predictive operations in buildings, homes, and organisations operating and managing commercial buildings. In building automation and management, a number of wired and wireless communication system are currently prominent (KNX, Modbus, M-Bus, etc.) often linked to building's operating system, in addition to numerous technologies of the building's users, which are completely decoupled from building's infrastructure. (such as safety, health, climate, etc.)

In a Smart Building perspective, a crucial point is that all information from such data streams can be captured and interpreted unambiguously and semantically correct as well as can be made available to other systems that need these information elements into their business process. This information flow shall be demonstrated in a building context. The same information elements shall be made accessible for processes and actors outside building contexts, such as energy, smart neighbourliness, health, and transport. Governments and regulatory agencies around the world are taking actions towards the energy efficiency of commercial buildings and residential neighbourhood, leveraging IoT technologies enabling new business opportunities and revealing benefits to all relevant stakeholders. More specifically these technologies are expected to enhance the functionality, capabilities, energy efficiency, and cost-effectiveness of buildings and homes [21].

Health Along with the rising ageing population with chronic diseases (such as people with hypertension, dementia, and obesity), healthcare becomes more and more difficult to manage due to insufficient and less effective healthcare services. E-health and assisted living for the elders and people with special needs can facilitate existing commercial equipment and sensors along with available communication channels to bring together end-users and their relatives, doctors, and help-centre assistance providers, in a decentralised and easy to integrate manner even for people not familiar with the technology. Healthcare applications are moving from hospital centred applications towards patient-centred applications.

By performing data analysis of health-related data, advanced health care services will provide users with similar data streams and user profiles recommendations about prevention and health condition improvements. Therefore, the elderly people and their relatives can obtain a better quality and independent life.

Transport Increasing demand for parking spaces in big cities indicates more search traffic, which leads to higher emission of CO_2, more congested traffic, and severer air pollution. Additionally, transportation and parking are two of the basic foundations of a smart city. By taking advantage of the parking data, cities will soon achieve better parking management and to capture new types of data and multiple analysis results [22].

In the transportation use cases, the services utilising prediction and optimisation algorithms can provide a balanced parking plan taking into account user preferences, parking history of vehicles (such as proper parking statistics), actual demand, etc. Pairing vehicle identity with a particular parking place will give rise to new business models such as variable prices based on parking place attractiveness, daily hours, or customer loyalty.

In order to address the needs of the individual residents, the management of parking space and proximity to access points can be tailored to user-defined profiles. Safety, predictability, reliability, accessibility, and comfort are elements that should be incorporated when implementing load balancing and resource administration of parking space and available areas. Access control and appraisal systems are functionality that needs to be supported. This is affected by what kind of user that wants to use the parking space. Visitors need to be kept separate from residents, but the needs of the user and preferred actions will have an impact on the recommended parking space/placement. Moreover, healthcare and blue light agencies must receive particular priorities.

Farming Along with the increasing population around the world, agricultural land is decreasing nowadays because of the industrial and modernisation process. The IoT-enabled advanced agricultural technologies are able to promote production efficiency, improve agricultural product quality and quantity, and reduce the production cost in the whole process of agricultural production to make the best use of the limited cultivated land [23]. Additionally, implementing precision farming/smart farming is the vision of future agriculture. Precision farming/smart farming can provide precise forecasting of supply and demand, real-time monitoring and resources management, and production quality maintenance during the entire life cycle of agricultural products [24]. By accurately sensing agricultural variations and scientifically using the strategy, farmers can economically improve the effectiveness of fertilisers and pesticides. Likewise, farmers can real-time monitor the health condition and needs of each animal and treat/feed them accordingly, thereby maintaining animals' health and nutrition.

4 Challenges in IoT

Today, the IoT is involved in all aspects of our lives with a strong impact on our working and living habits. However, there are lots of challenges that need to be addressed to make IoT a stable solution from a technical perspective and easily adaptable by the society. We distinguish two main categories: *technical* and *nontechnical* challenges; the latter becoming more and more dominant. However, in the end, both aspects are often tightly interwoven: nontechnical needs often can be seen as a technical challenge and vice versa.

4.1 Technical Challenges

From a technical perspective we extract here the most significant challenges for building an IoT system:

4.1.1   Interoperability from Technical Up to Semantic Level

In order to create an IoT in which machines communicate with each other, interoperability is a key issue. At the lower levels, existing standards, together with the merit of the communication stacks that permit us to exchange, e.g., a physical layer of the communication stack allows us easily to achieve some interoperability. However, once higher layers, e.g., the application layer, are needed, interoperability becomes an issue and a challenge.

Syntactic Interoperability The data structure and the format in which data is exchanged between heterogeneous IoT entities must be clearly defined. Data encoding on the sender side and decoding on the receiver side must follow compatible syntactic rules to avoid wrong interpretation of the message exchanged. We say that rules are compatible when the same grammar is used to define them.

Semantic Interoperability In the complex and heterogeneous IoT systems further means are required to create valuable semantic information out of raw data produced by "things" at the things layer. The semantic information is crucial in automatic management of services at the service layer.

Lack of Standards Standards can facilitate connections of new service providers and users, enable the interoperability between systems, subsystems, devices, and applications, as well as help devices and services to achieve better performances at a high level.

For more details on this topic, we refer a reader to Chap. 5.

4.1.2 Security and Privacy

To meet security requirements an IoT system must be able to provide services for data confidentiality, integrity, and availability. On the other side, the IoT devices can collect and save sensitive data or user's personal information such as names, addresses, dates of birth, pictures, card information, etc. These user's data can be stolen by a malicious attacker and maybe unauthorisedly used. Therefore, the measures should be taken in the development of IoT devices in order to protect the user's privacy.

These two very important topics are covered in detail in Chaps. 7 and 8 of this book.

4.1.3 Connectivity Challenges: Centralised vs. Decentralised Concepts

A straight-forward implementation of an IoT application could use a purely centralised approach: A server on which all data is saved, and that implements all services. However, for large scale systems where billions of devices are expected to be connected this approach faces a number of limitations:

1. *Scalability:* a centralised architecture creates a performance bottleneck, both in terms of computational resources and in terms of communication bandwidth.
2. *Reliability:* a centralised architecture creates a single point of failure; a failure might immediately affect all users. This holds in particular for the communication to all endpoints ("edges") for which it is virtually impossible to provide redundant communication links.
3. *Security and privacy:* in a centralised architecture all data and information are collected in a single point; a break of security or privacy will immediately have a huge impact.

Distributed architectures like edge-computing [25] strive to overcome the above issues by a more decentralised approach. The idea is to implement the higher-level functions closer to the objects. This can be done either in a local gateway under control of a user, or in a communication endpoint of the network operator. While edge-computing seems to provide solutions in particular for scalability and reliability and can reduce the impact of security and privacy issues, it introduces a number of new challenges:

- Concurrency issues such as races that occur if inputs appear at different points in time in different parts.
- Security issues can arise due to the fact that devices at the edges collecting a large amount of data are physically distributed. Here, it is of crucial importance that devices own proper security certificates that will establish secure connections to the edges.

- Network bandwidth. As more computation with a large amount of data is performed at the edges, bandwidth to the edges must be shifted to higher frequencies.
- Data handling. As more data is moved to the edges a new data handling strategy is required in order to comprehend all the data coming from the distributed edges.

Other promising solutions adopting decentralisation concepts are based on the use of peer-to-peer communications where actors with their devices exchange information directly without the involvement of a central server. Data stays locally and under the control of users. However, this can bring a new set of challenges, in particular from a security perspective. Beside companies as platform providers, device owners must also take necessary actions to deal with possible security attacks. *A recent study reported that more than 50% of device owners do not use any security tools to secure their devices. They even keep default passwords to protect their devices.*

4.2 Nontechnical Challenges

From a nontechnical perspective, there are also a couple of significant challenges that IoT systems are facing today:

1. *Legal and privacy issues.* Since many objects (users, sensors, devices, appliances, etc.) are connected, identified, communicated, monitored, and regulated in an IoT infrastructure, and efficient management of IoT data requires the gathering of big data that can be potentially misused to monitor users' habits breaching their privacy. Inappropriate implementation of the data gathering process can therefore raise serious legal and ethical issues. Furthermore, various software licenses of the IoT systems may raise policy issues in case of misuses.
2. *Customer awareness.* Most of the people owning IoT devices have little or no knowledge of what is happening with their data. According to a recent study by McAfee [26] with more than 6000 people surveyed globally even 33% of participants are concerned about how they can track and control what companies do with their data. This lack of awareness can have a negative impact on the fast adoption of IoT technology.
3. *Customer permission.* To implement an IoT application, a lot of sensors and devices will be installed or replaced in the existing systems and the customers need to change their habitual behaviour, for instance, to implement a smart building. Moreover, personal data might be collected, transferred, and analyzed. Therefore, how to convince the customers to install and to use the IoT application by clear and accessible privacy policies needs to be considered.
4. *Asset maintenance.* Asset maintenance issues of the equipped sensors/devices and lack of well-trained support staff are also the key challenges.

5   Conclusion

IoT is involved in every aspect of our daily lives. Its positive impact on our daily routines and the environment we live in is clearly visible through use cases in which IoT systems are applied. However, in order to bring IoT projects and their use cases to success, a number of challenges need to be addressed. Despite technical challenges, companies need to face a number of barriers coming from stakeholders including end-users. A tight communication between companies as IoT system builders and consumers of these systems is crucial. People need to be educated about positive aspects that IoT can bring to their lives but also risks that new technologies bring. According to [26] less than 40% of individuals use adequate identity protection solutions. A lack of awareness of what is happening in the background can lead to situations where all advantages of IoT are neglected by consumers' security and data privacy concerns.

In this book, all important IoT topics are covered through eight chapters. The particular focus is put on data security and privacy and technologies that can help users keep full control over their data. What exactly each chapter covers is described in the following.

6   Overview of the Book

Chapter 2 discusses technologies crucial for the implementation of *things* and *service management* layers of an IoT architecture. The first part of this chapter introduces how the real world can be sensed. After that, it gives a brief overview of IoT hardware and software platforms required to implement IoT gateways, crucial components for providing technical interoperability of heterogeneous objects. The next sections are dedicated to the technologies to connect devices together using protocols and the technologies for data storing and processing. Furthermore, the IoT platform developed in the VICINITY EU project and its features are introduced. Last but not least, the connected appliance design from electronics to the firmware is briefly introduced.

Chapter 3 is dedicated to the business layer of an IoT architecture. We present an overview of business models together with the impact that IoT has on them and the challenges identified. Moreover, we describe cross-domain use cases and value-added services (VAS) for the domains of smart buildings, energy, transportation, and health. The use cases that are developed and integrated into the VICINITY IoT platform are also presented as examples. Finally, an example of a digital transformation use case is given from Gorenje Group which uses comprehensive project portfolio management.

Chapter 4 discusses the methods and tools for IoT simulation and validation. We have introduced and compared the most known and commonly used IoT simulation tools. Two open IoT Test-beds which can be accessed by researchers to test the

IoT core components and prototypes are also described. Two testing scenarios of the VICINITY project are introduced as examples for the IoT simulation and Laboratory testing in the last section.

Chapter 5 introduces a reader to the technologies that deal with the interoperability of devices at the semantic level, one of the most significant IoT challenges. How this challenge can be resolved using the semantic platform developed within the EU Project VICINITY is demonstrated at the end of the chapter.

Chapter 6 gives a brief overview of standards that play one of central roles in enabling interoperability at technical, as well as, semantic level in complex IoT networks. Ongoing standardisation efforts and how we supported them bringing experiences and contributions from the VICINITY project are also presented here.

Focus of Chap. 7 is the security challenge and how it can be resolved. It discusses various security issues that must be considered and addressed during development of IoT platforms.

Chapter 8 focuses on data and privacy and what requirements need to be fulfilled that IoT solutions stay compliant with the GDPR regulations, officially published in May 2018. We show how we addressed this issue in the VICINITY project using homomorphic encryption and demonstrate the solution with hands-on code examples.

Chapter 9 helps readers dig into the IoT world and its benefits with the help of the VICINITY platform. This chapter serves a hands-on tutorial that illustrates how a reader can register his/her connected devices and connect it with another infrastructures and services using the VICINITY platform.

References

1. The internet gopher protocol (a distributed document search and retrieval protocol). https://tools.ietf.org/html/rfc1436
2. Tim, B.-L., & Robert, C. (1990). WorldWideWeb: Proposal for a HyperText Project.
3. The little-known story of the first IoT device. https://www.ibm.com/blogs/industries/little-known-story-first-iot-device/
4. Weiser, M. (Sept. 1991). The computer for the 21st century. *Scientific American.*
5. Ashton, K, et al. (2009). That 'Internet of Things' thing. *RFID Journal.* 22. Jg., Nr. 7, S. 97–114.
6. International Telecommunication Union. (2012). Overview of the Internet of Things. *Recommendation ITU-T Y.2060.* https://www.itu.int/rec/T-REC-Y.2060-201206-I
7. The Growth in Connected IoT Devices Is Expected to Generate 79.4ZB of Data in 2025, According to a New IDC Forecast. https://www.idc.com/getdoc.jsp?containerId=prUS45213219
8. Schelbey, Z., Hartke, K., & Bormann, C. (2014). RFC 7252 - The constrained application protocol (CoAP). In *Internet Engineering Task Force (IETF).*
9. Postel, J. (1980). DoD standard internet protocol. RFC 760, RFC Editor.
10. Postel, J. (1981). Transmission control protocol. RFC 793, RFC Editor.
11. Nielsen, H. F. (1994). The hypertext transfer protocol in the world-wide web. Master's Thesis, Aalborg University, Denmark, CERN, Geneva. twitter.com/frystyk.
12. Postel, J. (1980). User datagram protocol. RFC 768, RFC Editor.

13. Fielding, R., Gettys, J., Mogul, J., Frystyk, H., Masinter, L., Leach, P., & Berners-Lee, T. (1999). Hypertext transfer protocol–HTTP/1.1. https://www.hjp.at/doc/rfc/rfc2616.html.
14. Fielding, R. (2000). Architectural styles and the design of network-based software architectures. PhD Thesis, University of California, Irvine.
15. Crockford, D. (2006). The application/json media type for JavaScript object notation (json). RFC 4627, RFC Editor.
16. Zhu, Q., Wang, R., Chen, Q., Liu, Y., & Qin, W. (2010). IoT gateway: Bridging wireless sensor networks into Internet of Things. In *IEEE/IFIP International Conference on Embedded and Ubiquitous Computing* (pp. 347–352).
17. Chen, H., Jia, X., & Li, H. (2011). A brief introduction to IoT gateway. In *IET International Conference on Communication Technology and Application (ICCTA 2011)* (pp. 610–613).
18. Al-Fuqaha, A., Guizani, M., Mohammadi, M., Aledhari, M., & Ayyash, M. (2015). Internet of Things: A survey of enabling technologies, protocols and applications. *IEEE Communication Surveys & Tutorials, 17*(4), 2347–2376.
19. Silva, B. N., Khan, M., & Han K. (2018). Internet of Things: A comprehensive review of enabling technologies, architecture, and challenges. *IETE Technical Review*, 1–163, 5. Jg., Nr. 2, S. 205–220
20. Mynzhasova, A., Radojicic, C., Heinz, C., Kölsch, J., Grimm, C., Rico, J., et al. (2017). Drivers, standards and platforms for the IoT: Towards a digital vicinity. In *2017 Intelligent Systems Conference (IntelliSys)* (pp. 170–176).
21. Minoli, D., Sohraby, K., & Occhiogrosso, B. (2017). IoT considerations, requirements, and architectures for smart buildings-energy optimization and next-generation building management systems. *IEEE Internet of Things Journal, 4*(1), 269–283 (2017)
22. Guan, Y., Feng, W., Palacios-Garcia, E. J., Vásquez, J. C., & Guerrero, J. M. (2019). Vicinity platform-based load scheduling method by considering smart parking and smart appliance. In *2019 15th International Conference on Distributed Computing in Sensor Systems (DCOSS)*. Piscataway: IEEE.
23. Dagar, R., Som, S., & Khatri, S. K. (2018). Smart farming - IoT in agriculture. In *Proceedings of the International Conference on Inventive Research in Computing Applications (ICIRCA 2018)*.
24. Lee, M., Hwang, J., & Yoe, H. (2013). Agricultural production system based on IoT. In *2013 IEEE 16th International Conference on Computational Science and Engineering*. Piscataway: IEEE.
25. Ioannis Psaras. Decentralised Edge-Computing and IoT through Distributed Trust. In *MobiSys'18*, June, 2018.
26. McAfee. (2018). Key findings from our survey on identity theft, family safety and home network security. https://www.mcafee.com/blogs/consumer/key-findings-from-our-survey-on-identity-theft-family-safety-and-home-network-security/

Chapter 2
IoT Platforms

**Marie Madeleine Uwiringiyimana, Gomathi Nandagopal, Yajuan Guan,
Sašo Vinkovič, Johannes Kölsch, and Christopher Heinz**

1 Introduction

An IoT platform is a multi-layer technology that provides a set of ready-to-use features to speed up the development of IoT projects. It is an important piece in the IoT that enables the communication between objects.

IoT architecture is a system of numerous elements. But generally, a complete IoT system contains five components: hardware(sensors/devices), gateway, data processing in the cloud, connectivity (communication protocols) and user interface as shown in Fig. 2.1. Without the IoT platforms, there would be a gap between the hardware and application layer and it would be complicated to make everything work together. To fill this gap and solve that problem, there are different IoT platforms, which work as mediators between the varying kind of hardware and the different connectivity options. These platforms help developers in various tasks, such as handling various hardware and software communication protocols, providing security and authentication to devices and users as well as collecting, visualizing and analysing sensor data.

M. M. Uwiringiyimana (✉) · J. Kölsch · C. Heinz
TU Kaiserslautern, Kaiserslautern, Germany
e-mail: uwiringi@cs.uni-kl.de; koelsch@cs.uni-kl.de; heinz@cs.uni-kl.de

G. Nandagopal
Vel Tech Rangarajan Dr. Sagunthala R&D Institute of Science and Technology, Chennai, India
e-mail: gomathin@veltech.edu.in

Y. Guan
Department of Energy Technology, Aalborg University, Aalborg Øst, Denmark
e-mail: ygu@et.aau.dk

S. Vinkovič
Gorenje gospodinjski aparati d.o.o., Velenje, Slovenia
e-mail: saso.vinkovic@gorenje.com

© Springer Nature Switzerland AG 2021
C. Zivkovic et al. (eds.), *IoT Platforms, Use Cases, Privacy, and Business Models*,
https://doi.org/10.1007/978-3-030-45316-9_2

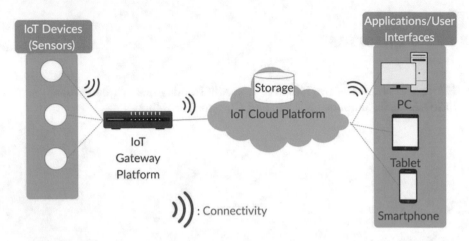

Fig. 2.1 Place of IoT gateway and cloud platform in IoT architecture

This chapter describes first how the real world can be sensed. Second, we present and evaluate the IoT gateway (hardware/software), IoT middleware and cloud platforms. The next sections will be about the technology to connect devices together using protocols. Furthermore, we discuss storing and processing data. Finally, we introduce the connected appliance design from electronics to firmware and we end up with the conclusion.

2   Sensing the Real World

Some inputs from the real world can be detected using a device called a "sensor". That means, a sensor allows us to interact with the surrounding environment. A sensor can measure a specific input/physical parameter, such as temperature, humidity, light, motion, heat, sound, etc. and convert it into a human-readable display signal.

Different types of sensors are used in everyday objects and are increasing rapidly in IoT. The smart devices such as refrigerators (see Fig. 2.2) are equipped with sensors, which can observe the events or changes in the environment and then data can be collected. The collected data could be, for example, the temperature, an image, a video and so on. These data provide us information continuously and allow us to recognize trends and make predictions.

To have an overview of the communication between sensors and applications, let us continue with an example of a smart refrigerator equipped with various sensors [2]. The described refrigerator has the ability to discriminate foods as well as to monitor their status. The main elements of the system architecture are sensor-equipped container, information server and maintenance application as shown in Fig. 2.3.

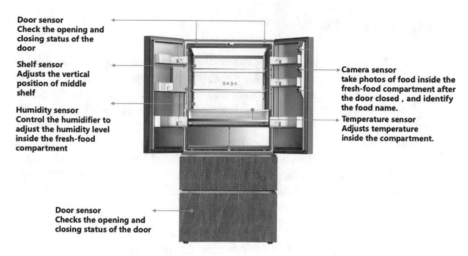

Fig. 2.2 Hisense Refrigerator's key sensors [1]

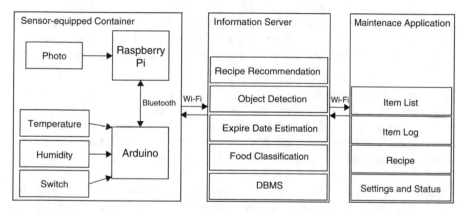

Fig. 2.3 System architecture of smart refrigerator [2]

Sensor-Equipped Container This is the hardware side. In order to detect the status inside the smart refrigerator, an Arduino, a Raspberry Pi and sensors (temperature sensor, humidity sensor and photosensor) are used. The temperature and humidity data are collected and the image can be taken.

Information Server The server has to connect the sensing hardware with the user interface/application. It receives the signals from sensors and then performs the required tasks.

Maintenance Application The user receives the processed data from the server via a mobile application on the smartphone or tablet. He/She can automatically get information from his/her connected refrigerator, although he/she is away from home.

3   IoT Gateway Platforms

A gateway is a bridge that allows two different environments to be able to communicate. In the Internet of Things, a gateway is a physical device or software program that serves as a connection point between the IoT devices and the cloud. Through the IoT gateway go all data/information moving from device to the cloud, or vice versa. It translates different low-level protocols from devices so that further transmission and processing over the Internet are possible. It plays the role of a bridge between IoT devices and the cloud. To build the gateway there are a variety of hardware and software platforms. The next subsections concentrate on the explanations of these IoT gateway platforms.

3.1   Evaluation of Gateway Hardware Platform

Physical IoT gateways are boards based on microcontrollers and microprocessors. The number of these boards is large and growing every day. For example, there are Banana Pro, Cubietruck, Raspberry Pi, BeagleBone, Pine A64+, SparkFun, Arduino, etc. The next paragraphs provide brief descriptions of some selected hardware boards.

Raspberry Pi 3 Model B   Raspberry Pi 3 Model B is a single-board computer equipped with BCM43438 chip that enables the communications over Wi-Fi 802.11n, Bluetooth Classic 4.1. The expansion header of this device has 40 pins that provide one UART connector, one I2C bus connector and two SPI bus connectors. Raspberry 3 Model B has four USB host ports (Fig. 2.4).

Banana Pro   Banana Pro is a single-board computer equipped with the AP6181 chip that enables Wi-Fi 802.11n communication. It does not have Bluetooth and ZigBee Modules. Banana Pro Expansion header has 40 pins which are used to

Fig. 2.4 Front side of Raspberry Pi 3 Model B board [3]

Fig. 2.5 Front side of
Banana Pro board [4]

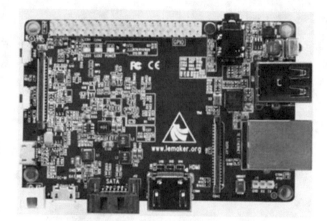

Fig. 2.6 Front side of Pine
A64+ board [5]

connect three universal asynchronous receiver-transmitter (UART) devices, two I2C
buses and one SPI bus. Pins 16, 17 are used in industry, automotive applications to
connect Controller Area Network (CAN). Two separate Universal Serial Bus (USB)
host ports and one USB On-The-Go (OTG) port are also available in the Banana Pi.
The banana pi pro diagram is shown in Fig. 2.5.

Pine A64+ Pine A64 single-board computer that supports different versions of
Android OS. Pine A64 board has two expansion headers in which the first one
consists of 40 pins and the second has 34 pins which is known as Euler bus. Pine A64
has one more additional expansion header that connects the Realtek RTL8723BS
based wireless module which enables support of Wi-Fi 802.11n and Bluetooth
Classic 4.0 technologies (Fig. 2.6).

Cubietruck Cubietruck single-board computer is equipped with an AP6210 wireless chip that enables the support of Wi-Fi 802.11n, Bluetooth Classic 4.0. Two
expansion headers are there in Cubietruck. 54 pins in the Cubietruck enable the three
universal asynchronous receiver-transmitter device connection, one I2C bus, SPI

Fig. 2.7 Front side of
Cubietruck board [6]

CubieBoard3 V1.0-0606 TSD

Fig. 2.8 Front side of Intel
Edison board [7]

bus. DVK570 expansion board enables to connect different modules to Cubietruck like Core2530 ZigBee module. Similar to the Banana Pro board, there are two separate USB host ports and one USB OTG port (Fig. 2.7).

Intel Edison The Intel Edison module is a system on a chip offered by Intel to produce IoT and wearable devices. This module enables communication over Wi-Fi and Bluetooth 4.0 LE (Fig. 2.8).

ESP8266 The ESP8266[1] is a low-power Wi-Fi microchip with full TCP/IP stack and microcontroller capability produced by manufacturer Espressif Systems in Shanghai, China.

ESP8266 WROOM Series include ESP-WROOM-02(D/U) and ESP-WROOM-S2.ESP-WROOM-02(D/U) is a low-power 32-bit MCU Wi-Fi module, based on the ESP8266 chip. This module integrates TCP/IP network stacks and includes

---

[1]ESP8266: https://www.espressif.com/en/products/hardware/esp8266ex/overview.

10-bit ADC with HSPI/UART/PWM/I2C/I2S interfaces. ESP-WROOM-S2 uses a 2 MB SPI flash connected to HSPI, working as SDIO/SPI slave, with the SPI speed being up to 8 Mbps. The ESP8266 Module can be integrated into space-constrained devices, due to its small size of 18 mm × 20 mm (ESP-WROOM-02)/16 mm × 23 mm (ESP-WROOM-S2).

Arduino Uno Wi-Fi The Arduino Uno Wi-Fi[2] has the same functions as the Arduino Uno Rev3 which is a microcontroller board based on the ATmega328P, but equipped with Wi-Fi and other improvements. It incorporates an 8-bit microprocessor from Microchip and has an on-board Inertial Measurement Unit (IMU). A device with Arduino Uno Wi-Fi can connect to a Wi-Fi network, using the secure ECC608 crypto chip accelerator.

The Wi-Fi Module is a self-contained system-on-a-chip (SoC) with an integrated TCP/IP protocol stack that can connect to a Wi-Fi network, or act as an access point. The Arduino Uno Wi-Fi has 14 digital input/output pins—5 can be used as PWM outputs—6 analog inputs, a USB connection, a power jack, an ICSP header and a reset button.

BeagleBoard BeagleBoard[3] provides low-cost single-board computers based on low-power Texas Instruments processors featuring the ARM Cortex-A series core. The designs are open source and components are available to manufacture compatible hardware. A serious product including PocketBeagle, BeagleBone, BeagleBone AI and BeagleBoard are provided by BeagleBoard.

SODAQ SODAQ[4] provides a series of SODAQ hardware products such as Autonomo, SODAQ SARA AFF N211 and Mbili. The projects based on SADAQ boards can be powered by small lithium batteries and solar panels alone, without the requirement for devices to be plugged into the wall.

The SODAQ SARA AFF N211 is a low power consumption developer board that is NB-IoT Supported and is compatible with Arduino. Moreover, a number of modules such as an integrated microcontroller (Microchip Atmel SAMD21), GPS (u-Blox M8Q for GPS, Galileo, GLONASS and BeiDou support), sensors (LSM303AGR digital magnetometer and accelerometer) are built in the board.

3.2 Evaluation of Gateway Software Platform

As mentioned in previous sections, there are not only the hardware platforms but also the software platforms to create IoT gateways. This section lists some open-source software platforms that can be used to build gateways.

---

[2] Arduino Uno Wi-Fi: http://store.arduino.cc/arduino-uno-wiFi-rev2.
[3] BeagleBoard: https://beagleboard.org/beagleboard/.
[4] SODAQ: https://shop.sodaq.com/sodaq-sara-aff-r410m.html.

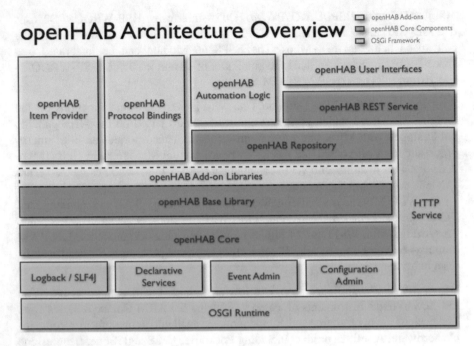

Fig. 2.9 OpenHAB architecture [8]

OpenHAB Platform One of the most popular solutions for smart homes is the open-source OpenHAB (open Home Automation Bus) platform. It is a Java-based platform developed with the objective to simplify home automation, but its features also make it useful in building IoT gateways. OpenHAB runs on the hardware, without the need for cloud service to work, keeps the data privately at home, speaks directly to the devices in the local whenever possible. The bundles of this based on the OSGi (Open Services Gateway initiative) platform are deployed on an Equinox runtime. Figure 2.9 shows an OpenHAB architecture overview.

DeviceHive Platform DeviceHive is an open-source platform for M2M communication which provides all means for building IoT, especially IoT gateways. Unlike OpenHAB, it does not use OSGi and its architecture based on D-BUS as shown in Fig. 2.10. DeviceHive is a scalable, hardware-based cloud-agnostic microservice-dependent platform along with device-management APIs in various protocols that allows to set up and monitor the connectivity of the device, control them and also to analyse their behaviour.

OpenRemote Platform OpenRemote is an open-source platform developed in 2009 with the objective of overcoming the problems caused by attempts at integration between different protocols and already existing M2M communication solutions. It integrates different protocols and solutions that are available for smart building, smart city automation and it also offers visualization. Figure 2.11 illustrates the OpenRemote architecture. It can be split into three main components:

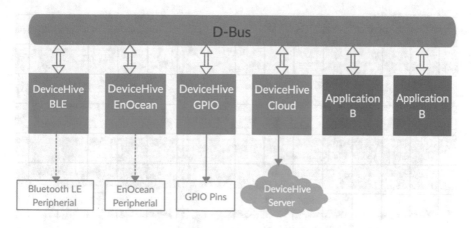

Fig. 2.10 DeviceHive architecture overview [9]

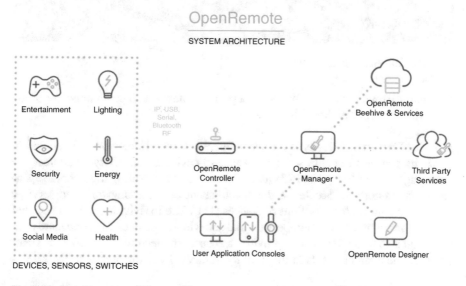

Fig. 2.11 OpenRemote architecture [8]

cloud-based configuration tools called OpenRemote Designer, local runtime controller and control panels (apps).

AllJoyn Platform AllJoyn is an open-source platform created in 2013 to enable seamless device interoperability. "AllJoyn is an open-source software framework that makes it easy for devices and apps to discover and communicate with each other. Developers can write applications for interoperability regardless of transport layer, manufacturer, and without the need for Internet access. This framework can be used on all modern operating systems since it offers an abstraction layer for Android, IOS, Linux, and Windows" [8]. For AllJoyn, the cloud connection is

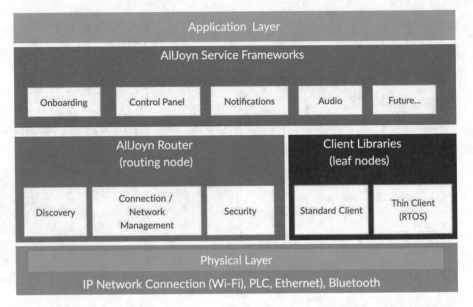

Fig. 2.12 AllJoyn architecture [10]

optional. AllJoyn is flexible, promotes a proximal network. The architecture of the AllJoyn framework is shown in Fig. 2.12.

IoTivity Platform IoTivity is an open-source software framework that started in 2015. Its goal is to enable seamless device-to-device connectivity to address the emerging needs of IoT. "The IoTivity project was created to bring together the open-source community to accelerate the development of the framework and services required to connect these billions of devices [11]". The IoTivity architecture created a new standard by which a number of wired and wireless devices can connect to each other and also to the internet. It provides a robust architecture that works for smart devices. The main elements of the IoTivity framework are illustrated in Fig. 2.13.

Eclipse Kura Platform Eclipse Kura is an eclipse IoT project that offers a Java/OSGi (Open Services Gateway initiative)-based platform for building IoT gateways. This platform runs on top of Java Virtual Machine (JVM) and leverages OSGi, a dynamic component for Java.

Kura brings the following advantages:

- Simplification of the process of writing reusable software building blocks
- Simplification of the management of network configurations, the communication with IoT servers as well as the remote management of the gateway
- Kura offers APIs access to the hardware gateways interfaces (GPIOs, I2C, Serial ports, etc.). It has also MODBUS, CAN bus APIs, etc.

Figure 2.14 illustrates an overview of Eclipse Kura platform.

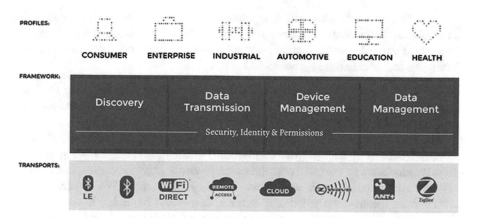

Fig. 2.13 IoTivity architecture [11]

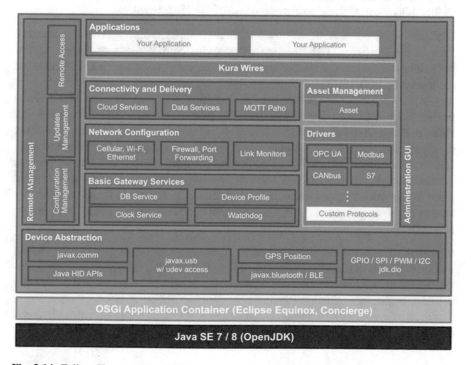

Fig. 2.14 Eclipse Kura overview [12]

4   IoT Middleware

IoT middleware works as a bridge to interconnect the heterogeneous components and to facilitate the adaptation, resource identification and management, data management, knowledge extraction, privacy and security developments. This section lists some IoT Middleware Frameworks.

Hydra   The Hydra project[5]—Networked Embedded System middleware for heterogeneous components in a distributed architecture has led the P2P network to provide interoperability and security for IoT applications.

The developers are able to interconnect various types of physical devices into their applications by using Hydra middleware, thereby providing easy-to-use web service interfaces to control heterogeneous physical objects regardless of their network technologies, such as Zwave, Wi-Fi, ZigBee, LoRaWAN, etc. Hydra incorporates means for Device and Service Discovery, Semantic Model Driven Architecture, P2P communication, and Diagnostics. Hydra-based physical objects and services are secure and trustworthy through distributed security and social trust components of the middleware. The new name of the Hydra middleware is changed to the "LinkSmart[6]" middleware at the end of the Hydra project.

Ubiware   Ubiware is an agent-based middleware that assigns a proactive agent to each of the resources. The Ubiware agent architecture consists of three layers: a behaviour engine layer, a middle layer and a resources layer which are represented as Java components [13].

Ubiware employs the Distributed AI, Semantic Web and Human-Centric Computing technologies to the Ubiquitous Computing domain. The developers are able to create the self-managed complex industrial systems including mobile, distributed, heterogeneous, shared and reusable components of different nature. Ubiware is a domain-independent middleware platform, thereby offering support for interconnection, interoperability, communication, interaction, self-awareness, and planning for various types of resources, systems and devices [14]. Interoperability among the objects requires the use of metadata and Ontologies, which form the core of the Ubiware.

OpenIoT   The OpenIoT[7] middleware infrastructure supports flexible configuration and deployment of algorithms for collection, and filtering information flows transferring from the internet-connected components, meanwhile creating and processing valuable business/application events.

OpenIoT enables a new series of open large scale smart IoT applications in terms of a utility cloud computing delivery model.

---

[5]Hydra Project: https://vicinity2020.eu/vicinity/content/hydra.

[6]LinkSmart: https://www.linksmart.dk/news.php.

[7]OpenIoT: https://github.com/OpenIotOrg/openiot.

OpenIoT provides access to additional and cumulatively IoT-based resources and technologies. Among others, OpenIoT researches and offers the means for formulating and managing environments including IoT resources, which is able to deliver on-demand utility IoT services such as sensing as a service.

OpenIoT is relevant to a number of pertinent scientific and technological fields consisting of:

- Middleware for sensors and sensor networks,
- Ontologies, semantic models and annotations for representing internet-connected objects, together with semantic open-linked data techniques,
- Cloud/Utility computing, comprising utility-based security and privacy schemes.

FIWARE FIWARE[8] is an open architecture and operative software for the generation and delivery of services, correlated with various areas performed in the context of the FI-PPP program (Future Internet Public-Private Partnership). The FIWARE platform aims to establish an open sustainable ecosystem around public, royalty-free and implementation-driven software platform standards that can facilitate the development of novel intelligent applications in different domains [8].

FIWARE platform is supported by the FIWARE community, all people who support FIWARE materialized through the FIWARE Foundation, the legal entity to support FIWARE community and the FIWARE OSC, the Open-Source Community of persons who develop the FIWARE technologies.

5   IoT Cloud Platforms

Larger cloud providers offer their cloud business into the IoT. Different solutions have been offered like Infrastructure-as-a-Service (IaaS) backend which provides hosting space and processing power for services and applications. The backend is streamlined for applications which have been updated and integrated into IoT platforms like VICINITY Platform, Amazon Web Services, Microsoft Azure, Google Cloud IoT platform. Although these platforms offer almost similar properties, there are certain differences which are helping them to keep hold of their positions.

VICINITY Platform VICINITY project,[9] funded by the European Commission's Directorate-General for Research and Innovation (DG RTD), under its Horizon 2020 Research and Innovation Programme (H2020) starts in 2016. VICINITY platform is an open virtual neighbourhood network to connect IoT infrastructures and smart objects [15]. It is a decentralized platform, which resembles a "social network" (known as "virtual neighbourhood" in VICINITY) and provides "interoperability as a service" for infrastructures in the Internet of Things. Through the VICINITY

---

[8]FIWARE: https://www.fiware.org/.

[9]VICINITY Project—https://www.vicinity2020.eu/vicinity/.

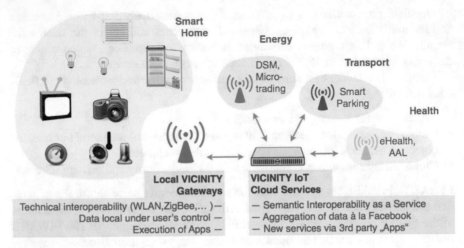

Fig. 2.15 VICINITY overview [8]

platform, different IoT ecosystems can be interconnected. That means it allows interaction with smart devices from other ecosystems as if they were their own. The user is able to control his shared smart devices and data by setting access rights at the neighbourhood manager. VICINITY platform can be used in various areas. Some use cases are smart home, smart energy, smart transport and eHealth (Fig. 2.15).

Amazon Web Services Platform Amazon Web Services, shortly called as AWS, is a cloud platform launched in 2006 by Amazon, which allows the device-to-device and device-to-cloud connection. It supports HTTP, MQTT and WebSocket and provides authentication/encryption to secure data. As illustrated in Fig. 2.16, the AWS platform key features are AWS IoT Device SDK, Device Gateway, Authentication and Authorization, Registry, Device Shadows and Rules Engine.

Key features [17]:

- Scalability
- Privacy and Security
- Pay-as-you-go model
- High availability and flexibility
- Data analytics and storage of high volume data
- On-demand services like other platforms
- Provides Hardware resources

Microsoft Azure Platform Microsoft Azure is a cloud computing platform, launched by Microsoft in 2010. This platform helps in service management, service hosting, data storage and development. The three main types of clouds in Microsoft Azure (Fig. 2.17) are IaaS (Infrastructure as a Service), PaaS (Platform as a Service) and SaaS (Software as a Service).

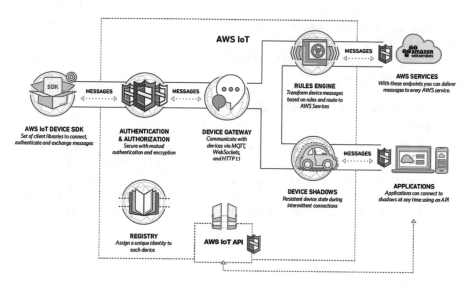

Fig. 2.16 Amazon Web Services overview [16]

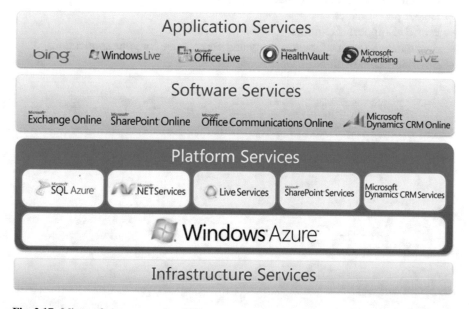

Fig. 2.17 Microsoft Azure overview [18]

Key features [17]:

- Build on what already established
- Get more benefits from your existing assets
- Small changes and big reflection
- Trusted support

- Expertise in development to deployment
- Connectivity of any device
- Skilled partners and powerful innovation
- Data insights
- Scalability
- Easy way for business transformation
- Agility

Google Cloud Platform Google Cloud Platform (GCP) is a platform offered by Google in 2008, which enables to build, deploy and scale applications, websites and services on the same infrastructure as Google. The data lifecycle from initial acquisition to final visualization has four steps: ingest, pipelines, storage and analytics, application and presentation. Figure 2.18 shows the real-time stream processing of GCP.

Key features [17]:

- Run on Google's infrastructure
- Scalability
- Compute, storage and services
- Higher performance
- Provided support if required
- Assurance of Google Grade security and compliance for your applications
- Environment safe cloud

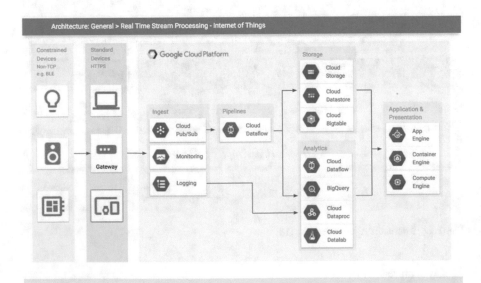

Fig. 2.18 Google Cloud Platform architecture [19]

6   Selection of IoT Platform and Building Your Own Smart Device

6.1   Selection of IoT Platform

From the above sections, every platform has at least one better specific thing to offer in the Internet of Things. Therefore, it is not easy to choose the right platform. The following advice could help you to select the right IoT platforms for your project. First, you need to get a detailed comparison of different IoT platform's features. Second, you need to list the specific requirements for your smart devices. Once you have listed the requirements, you can use available IoT platforms that meet your requirements.

In addition to the comparison already presented, the following criteria may also be considered.

Hardware Platform Selection The main prerequisites to the IoT gateway hardware platform may be high performance, connectivity and autonomous operation.

Software Platform Selection When choosing the software platform, the criteria can be the available IoT data collection protocols, security, portability, extendability, developer documentation, and examples and support community.

Cloud Platform Selection For IoT cloud platform selection, some of the main criteria are:

- It should allow to pick a platform with extensive protocol support for data ingestion.
- It also should ensure the platform has a robust capability for offline functionality.
- It also should make sure that the platform provides cloud-based orchestration to support device lifecycle management.
- It also should provide a hardware-agnostic scalable architecture.
- It also should provide a comprehensive analytics and visualization tools to make a big difference.

6.2   Building Your Own Smart Device Using VICINITY Platform

After analysing various IoT platforms and their features, let us now build a project. This project uses a Raspberry Pi, Eclipse Kura and VICINITY platform to control Philips Hue light from anywhere. As shown in Fig. 2.19, the main required components are:

- Philips Hue bulb [20]
- Philips Hue bridge [20]

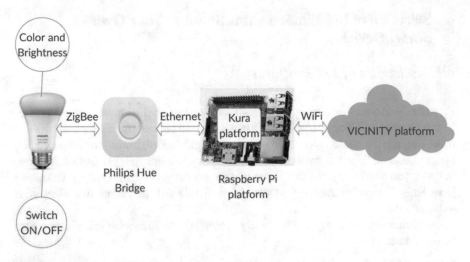

Fig. 2.19 Philips Hue lights in VICINITY platform

- Raspberry Pi 3 Model B as a gateway hardware platform [3]
- Eclipse Kura as gateway software platform
- VICINITY platform as cloud platform

To control the Hue light, an IoT gateway is created using Eclipse Kura and Raspberry pi. Kura platform is installed on the Raspberry Pi server and java code is in Eclipse IDE written. With the Raspberry Pi commands would the light bulb be available in the VICINITY Neighbourhood manager. Therefore, from anywhere, you can change the light colour and brightness of the Hue bulb. It is also possible to switch ON or OFF the light. For testing the project, the HTTP methods from Postman (a popular tool in API testing) can be used.

7 Connecting Devices

As IoT based on connectivity, communication protocols play an important role in enabling intelligent objects to send and receive data. The connectivity manages the data movement between IoT endpoints (sensors and gateways) and cloud. There are a variety of IoT protocols not only the wired communication protocols but also wireless communication protocols. The following paragraphs present some of them.

7.1 Device Connectivity Protocols

A protocol overview of IoT devices and applications will clarify the IoT layer technology stack [21]. The IoT encompasses a wide range of industries that scale from a single device up to the abundant cross-platform implementation of embedded technologies, cloud systems communicating in real-time.

List of protocols and standards that help to power IoT devices and applications drill down on specific layers.

Instead of trying to fit all IoT Protocols on top of OSI Model, the protocols can be broken into the following layers to provide a certain level of organization: Infrastructure (ex: 6LowPAN, IPv4/IPv6, RPL), Identification (ex: EPC, uCode, IPv6, URIs), Comms/Transport (ex: Wi-Fi, Bluetooth, LPWAN), Discovery (ex: Physical Web, mDNS, DNS-SD), Data Protocols (ex: MQTT, CoAP, AMQP, Websocket, Node), Device Management (ex: TR-069, OMA-DM), Semantic (ex: JSON-LD, Web Thing Model), Multi-layer Frameworks (ex: AllJoyn, IoTivity, Weave, HomeKit).

Generic Protocols—Wi-Fi, Cellular, Ethernet The IEEE 802 Standard embraces a group of networking standards that enclose the physical layer specifications of technologies right from Ethernet to wireless. IEEE 802 has been divided into 22 parts that encompass the physical and data-link aspects of networking. The known specifications are 802.3 Ethernet, 802.11 Wi-Fi, 802.15 Bluetooth/ZigBee and 802.16.

The specifications of 802.11 utilize the Ethernet protocol and Carrier Sense Multiple Access with Collision Avoidance (CSMA/CA) for sharing the path. Phase-shift keying (PSK) was the original modulation used in 802.11. Complementary code keying (CCK) is being used in a few newer specifications. Higher data speed, reduced vulnerability to interference are being provided by modulation methods.

The cellular network is a communication network in which the last link is wireless. The network has been distributed over land areas called "cells", where each cell is served by at least one fixed-location transceiver. The fixed-location transceiver is also known as the base station that provides the cell with the network coverage which is utilized for transmitting voice, data, and other kinds of content. Cell generally utilizes varying sets of frequencies from the neighbouring cells, to avoid interference and it also provides guaranteed service quality within each cell.

When cells are joined together, they provide radio coverage over a wide geographical area. This allows numerous portable transceivers (e.g., mobile phones equipped with mobile broadband modems,) to communicate with each and every device in the network, through base stations, though the transceivers are moving via more than one cell during transmission.

LoRaWAN LoRaWAN is low-power wide area network (LPWAN) based on spread spectrum modulation techniques and it is implemented in the Internet of Things (IoT) networks worldwide for energy management, natural resource reduction, pollution control, infrastructure efficiency, disaster prevention, etc.

Bluetooth It is a wireless technology used to transfer data between fixed and mobile devices in short distances through short-wavelength UHF radio waves in the industrial radio bands to build personal area networks (PANs). It is designed as a wireless alternative to RS-232 data cables. Bluetooth uses frequency-hopping spread spectrum and it divides transmitted data into packets. Each packet is transmitted on one of the 79 designated Bluetooth channels. Each and every channel possess bandwidth of 1 MHz. 1600 hops are performed per second, with adaptive frequency-hopping enabled.

Zigbee Zigbee is a high-level communication protocol used to create personal area networks such as home automation, medical device data collection and other low-power low-bandwidth needs. Zigbee specification technology is simpler and less expensive than other WPANs such as Bluetooth and Wi-Fi. Zigbee Applications are wireless light switches, home energy monitors, traffic management systems and other consumer and industrial equipment. Zigbee is used in low data rate applications that require long battery life and secure networking and it is intended for intermittent data transmissions from a sensor or input device.

LTE-M NgIoT LTE-M which is an enhanced Machine-Type Communication is a type of low-power wide area network designed for machine-to-machine and Internet of Things applications.

The LTE-M has a higher data rate, mobility and voice over the internet than NB-IoT network. NGIoT: NGIoT is Next Generation IoT will provide human-centric digital transformation, in both the private and public sectors by including fields like Edge Computing, 5G, Artificial Intelligence and Analytics, Augmented Reality and Tactile Internet, Digital Twin and Distributed Ledgers.

Narrowband IoT (NB-IoT) Mobile Telephony Protocols were developed by the 3rd Generation Partnership Project (3GPP) [22]. Wider sort of cellular services were enabled by 3GPP. 3GPP developed Narrowband Internet of Things (NB-IoT). NB-IoT is a low-power wide area network (LPWAN) radio technology standard. eMTC (enhanced Machine-Type Communication) and EC-GSM-IoT were also developed by 3GPP. Indoor coverage, long battery life was the focus of NB-IoT. A bandwidth of 200 kHz is used by NB-IoT. NB-IoT for the sake of downlink communication used OFDM. NB-IoT also used SC-FDMA for the uplink communication.

8   Storing and Processing of Data

IoT creates, stores and processes content across many physical locations. To ensure this data is adequately secured, collated and processed is a challenging task. With IoT, the challenge of putting arms around all of the content delivered by the devices is much greater. In many instances, the value of the data may not be best served by storing the entire content. For example, a camera that counts cars passing a traffic intersection does not need to store the entire video, just report back the number of cars counted over specific time periods. The video data could be moved back at some time in the future or simply discarded.

The next point to consider is the timely processing of data. IoT devices may need to make local processing decisions quickly and not tolerate the latency of reading and writing the data into a core data centre for processing to occur.

The distributed data from IoT devices and processing the device data requirement means that businesses need to add the capability to push compute and applications to the edge and, in many cases, pre-process data before it is uploaded to the core data centre for long-term processing.

8.1   IoT Platform Supporting Connectivity

An IoT platform is a tool to develop and run an application. In IoT for hardware vendors, the platform is embedded board to write the IoT applications. For cloud service providers, the platform is where the developer by consuming the data delivered by the sensor creates an application. In general, the IoT platform is the middleware layer for consuming data from sensors and devices to produce reasonable actions based on that insight. IoT platform offers well-defined APIs by which developers can connect any hardware platform and they can consume the cloud-based services.

Centralized Platforms An IoT platform as a service (PaaS) with centralized architecture is a hub (typically powered by the cloud) that controls the execution of nodes (smart devices). But the centralized platform is not sufficient to implement industrial IoT solutions. PaaS is a framework where developers can build upon to develop the applications. PaaS helps the development and testing of applications very quickly, in a cost-effective manner. With PaaS technology, third-party providers can manage OSs and virtualization.

PaaS allows creating applications using the software components which are built into PaaS. Applications using the PaaS inherit the cloud characteristics which are scalability, high-availability, multi-tenancy. Enterprises are benefitted from PaaS because it decreases the amount of necessary coding, automates the business policy.

Decentralized Platforms In decentralized platforms, nodes in an IoT interact without the control of a central authority framing a fundamental basis to enable the next generation of IoT enterprise. Decentralized IoT architecture along with the logic moved to (network) edge will offer welfare to IoT platforms. Five features the decentralized IoT architecture offer are multi-network approach, scalable and interoperable implementation, low-power consumption, intuitive data and device management and Artificial Intelligence at the Edge.

8.2   IoT Data Storages

A lot of open-source platforms are available to collect and store sensor data to the cloud. Few platforms like ThingSpeak provide the app to analyse and visualize the data in Matlab. Arduino, Raspberry Pi, BeagleBone are used to send sensor data. A separate channel can also be created to store data. Other IoT platforms available in the market are ThingWorx 8 IoT Platform, Microsoft Azure IoT Suite, Google Cloud IoT Platform, IBM Watson IoT Platform, AWS IoT Platform, Cisco IoT Cloud Connect, Salesforce IoT Cloud, Kaa IoT Platform, Oracle IoT Platform. Managing and storing the IoT data is a challenging task. For example, plant machinery that was run by the server, collection of data from a remote branch office is very difficult. One and only option is storing the data outside the data centre which is nothing but the edge. In edge computing, managing and computing is done outside the data centre.

Edge computing is in demand in recent years due to the large volume of data being generated in non-core datacentre locations. IT centres should ensure the secured storing and processing of IoT data. To process the data in a safe manner many factors should be taken into account which are discussed below. The first factor is a large amount of money should be invested in external networking. Second, there is no need to store the entire device data, instead data aggregation alone is enough. The third need is to process the data instantly.

All the above-mentioned factors are not possible if we store the entire content of IoT data in data centres. Therefore to manage the IoT data, information Life cycle Management (ILM) has to be introduced. ILM does not deal only with cost optimization of data storage but it also ensures that the data is kept in the right location for instant processing in order to get business insights from it by means of machine learning and AI algorithms.

Let us discuss what are the provisions available in IT sectors for storing and processing the IoT data in the present scenario.

- Public cloud not only provides storage but it also allows people to use AI tool to process and analyse the data.
- Supplier solutions in IoT, like Amazon Web Services (AWS), uses Snowball appliance. Snowball is a server with storage that is typically used to physically transfer and process the data locally from an offsite location.
- Google Cloud Platform (GCP) provides a cloud platform to process the unstructured IoT data.
- DDN has also developed converged infrastructure to store and process the data with the support of on-premise ML/AI systems.
- A product has been developed by Excelero for scalable file storage and low latency analytics requirements.

9 Connected Appliance Design from Electronics to Firmware

9.1 Introduction

The internet technology stands on the brink of a new era of ubiquitous computing and communication with the large explosion of the IoT and its enabling applications and services. In the last few years, the computational power of the devices has been increased, but simultaneously their size has been decreased dramatically, resulting in their "disappearance" and seamless integration in larger systems that are able to control the environment (e.g., network of household appliances).

Electronics of household appliance is an embedded device equipped with sensors, processors and actuators that guarantee a real-time performance of the device. Most of these appliances have quite strict requirements for reliability and guaranteed performance due to their safety-critical nature and the fact that any malfunction could have catastrophic consequences to the lives of people and the environment in general.

Nowadays most household appliances comprise the microcontrollers, but not wireless communication interfaces; therefore, many of them still cannot communicate with the Internet. Recent advances in embedded computing foster household appliances to support wireless connectivity, thus the penetration of the connected white goods in the field of the IoT is expected in the next years.

9.2   Requirements

When it comes to efficient and reliable design of the IoT household appliance or any other device, electronics and the communication module must be considered as the main component in addition to suitable appliance construction design due to efficiency and perception of reliable radiofrequency signal strength. In general, the software plays a fundamental role in designing the IoT appliance, different levels of software development must be linked properly. This usually reflects enormous challenges for the development teams particularly for the organizations that are about to change the years of the conventional paradigm of development into a service-oriented approach (Fig. 2.20).

The hardware part of connected or regular non-connected appliance consists of a user interface that allows for control and monitoring of appliance. It is further coupled with the power board that usually takes control over peripheral components such as heating and actuating elements. There are some appliances where their operation is handled by more than two electronics; however, the control board is the most common electronics that hosts the communication module. The main requirements of hosting electronics are a physical interface to connect the communication module (e.g., UART) and the user interface with additional features to configure the network parameters of the module.

On the other hand, the software design of connected appliance requires a broader understanding of the entire chain of events, this is from appliance to application. If designed properly, hardware does not require maintenance and will not generate additional costs, but on the contrary, the maintenance of software is the crucial element of the connected appliance and could reflect enormous unforeseen costs if not managed properly. In addition to basic source code implementation connected appliance requires also the representation of the functional profile, which is a dictionary of functionalities, and implementation of communication protocol that

Fig. 2.20 The elements that compose the connected appliance

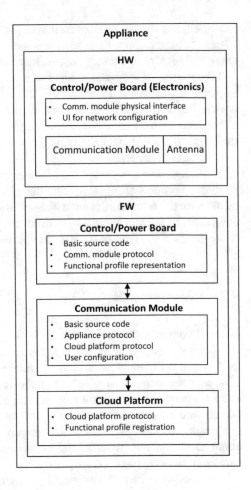

links the host electronics and communication module. There are two protocol implementations at the communication module, the first connects the module with the host electronics and the second with the cloud platform that allows for registration of appliances based on its functional profile.

9.3 Electronics

9.3.1 Definition of Functional Profile

An adequate structure is required (e.g., mark-up language) to define the functional profile of appliance which is actually a functionality dictionary with built-in features of an embedded device. The individual functionality of a particular device consists of the following properties [23]:

Table 2.1 Refrigerator functional profile

Functionality	Identifier	Access	Type	Valid values
DEVICE_STATUS	2010	Read-only	Enumeration	2011 = IDLE 2012 = SERVICE
TEMPERATURE	2070	Read-only	Signed Integer	This value is read from the NTC sensor
DOOR_STATUS	2090	Read-only	Enumeration	2091 = CLOSED 2092 = OPEN
LIGHT_STATUS	2130	Read-write	Enumeration	OFF = 2131 ON = 2132
ALARM	2160	Read-only	Enumeration	2161 = NONE 2162 = DOOR_OPEN
COMPRESSOR	2330	Read-write	Enumeration	OFF = 2331 ON = 2332

- name of the functionality,
- functionality identifier,
- type of value,
- type of access to the value,
- automatic synchronization,
- initial value,
- range of valid value.

Table 2.1 depicts the extract of the functional profile which is applicable for a cooling category of household appliances.

9.3.2 Hardware Design

To keep the balance between the final price of the product and the unification to apply the solution on different categories and models of household appliances, designing the hardware or electronics of the appliance could be an enormous challenge nowadays. The following requirements must be met:

- the electronics' interface to connect the communication module,
- the location of communication module's antenna and
- the user interface add-on to configure the communication module.

9.3.3 Firmware Design

It is recommended that the connected code does not interfere with the basic source code of an appliance and should work as a wrapper. There are three main pillars to consider:

- protocol to connect electronics and communication module,

- functional profile and internal protocol implementation to gather the information from multiple electronics and sensors,
- firmware update either wired or over the air solution, but could be a challenge due to the diversity of different electronics.

9.4 Communication Module

9.4.1 Security Aspect

The data that is generated by the appliance's electronics, in particular the communication between electronics and the communication module, should be handled with a special care due to safety concerns and potential security breaches. It is highly recommended that the content is properly encrypted. A tailored microcontroller and curtailed memory capacity are usually considered to keep the costs down of the final product; therefore, adding a custom made implementation of encryption protocol could be despite the importance of a big challenge.

9.4.2 Operating Temperature and Dimensions

In the field of household appliances development the operating temperature is not an issue generally, but in some categories such as cooking appliances, the temperature is a quite important parameter (e.g., oven and hob). The tendency is to provide the module that dimensions are as smallest as possible, otherwise the size to assure the cross-category unification could have an impact on mounting in some cases. Also, due to unification reasons, the module should be certified at the operating temperature of at least 85 °C.

9.4.3 Testing and Certification

It is highly recommended to run as much as possible testing on the communication module (exposure to high temperature, humidity and vibration) as the consequences at the market in case of failure could be detrimental. The certification goes along the testing and longer the list of certificates better are chances to succeed on the market.

9.4.4 Firmware Development

The firmware of communication module is the bridge between the physical appliance and the external world. It is of a pivotal role to design it properly by taking into consideration the extensibility and maintenance. The blocks to consider are as follows:

- protocol to connect electronics and communication module,
- protocol to connect communication module and cloud platform and
- a friendly way to configure the module either over website or application.

9.4.5 Power Consumption and Power Management

Today the power consumption of electronics itself is generally a big concern of manufacturers due to various legislative and regulative demands. The communication module is additionally required to apply connectivity functionalities which customers require more and more frequently. Such requirements further increase power consumption and impacts the concern of the manufacturers.

The power management protocol could mitigate the increase of power consumption; therefore, the consideration to define the protocol at the very beginning of the project is strongly advised. The idea is to switch the module off at the certain times and switch it back either periodically (e.g., remote reception of messages delivered by cloud or application) or upon control panel activity (e.g., user operates with the appliance by pressing the control buttons).

9.4.6 Mounting and Antenna Positioning

A modular approach with only one communication module is the promising solution to keep the maintenance costs down and to unify the way how the procurement and the stock keeping units are handled.

There are three options when it comes to mounting the module with the appliance's host electronics:

(a) directly via connectors,
(b) directly by soldering and
(c) indirectly by using mounting case and connecting cable.

The first option provides the portability between different appliances, but antenna position is limited to the position of electronics that hosts the module. The second one is reasonable for higher demands and big manufacturing volumes; however, the after-sales replacement will not be possible in case of malfunction. In appliances with a limited mounting space, a third option with the connecting cable could be considered.

To gain the best possible signal strength it is of utmost importance to position the module's antenna properly. It could be a challenge to identify the best possible position due to design constraints of the appliance (e.g., standalone vs. built-in refrigerator, built-in oven with glass vs. metallic control panel). In principle, an external antenna could be applied but due to additional costs of cabling and connector, a solution with an on-board antenna should be always a preferential choice.

10   Conclusions

In this chapter, we have presented the IoT gateway, middleware and cloud platforms. First, we have defined selected platforms to show the differences between them. We have also explained how to select the IoT platform to create your own smart device. Overall, each project is unique and every platform has something better to offer. Therefore, choosing the right platform depends on the need of your project. In addition, we have discussed connecting devices using communication protocols. Protocols that are to be used to manage the data movement in IoT infrastructure have been defined. Moreover, we have presented the data storing and processing in the IoT. The last section of this chapter was about connected appliance design from electronics to firmware.

References

1. Hisense. http://global.hisense.com/
2. Kwon, T., Park, E., & Chang, H. (2016). Smart refrigerator for healthcare using food image classification. In *Proceedings of the 7th ACM International Conference on Bioinformatics, Computational Biology, and Health Informatics - BCB '16*, Seattle, WA (pp. 483–484). New York: ACM Press.
3. Kurniawan, A. (2016). *Getting Started with Windows 10 IoT Core for Raspberry Pi 3*. Riverside: PE Press.
4. Follmann, R., & Zhang, T. (2015). *Banana Pro Blueprints*. Birmingham: Packt Publishing Ltd.
5. PINE A64 (+). https://www.pine64.org/devices/single-board-computers/pine-a64/
6. Cubietruck | CubieBoard. http://cubieboard.org/tag/cubietruck/
7. Arduino - IntelEdison. https://www.arduino.cc/en/ArduinoCertified/IntelEdison
8. Heinz, C., García-Castro, R., Sveen, F., et al. (2016). Analysis of Standardisation Context and Recommendations for Standards Involvement. https://vicinity2020.eu/vicinity/sites/default/files/documents/vicinity_d2.1_analysis_of_standardisation_context_and_recommendations_for_standards_involvement.pdf
9. IoT Toolkit Overview. https://docs.devicehive.com/v2.0/docs/iot-toolkit-overview
10. Villari, M., Celesti, A., Fazio, M., & Puliafito, A. (2014). AllJoyn lambda: An architecture for the management of smart environments in IoT. In *2014 International Conference on Smart Computing Workshops* (pp. 9–14). Piscataway: IEEE.
11. About | IoTivity. https://iotivity.org/about
12. Kura Wires Overview. https://eclipse.github.io/kura/wires/kura-wires-intro.html
13. Razzaque, M. A., Milojevic-Jevric, M., Palade, A., & Clarke, S. (2016). Middleware for Internet of Things: A survey. *IEEE Internet of Things Journal, 3*(1), 70–95.
14. Nikitin S., & Lappalainen, M. (2010). Tekes project proposal: SOFIA Full title: Seamless Operation of Forest Industry Applications. http://www.cs.jyu.fi/ai/OntoGroup/SOFIA/SOFIA_old.pdf
15. Vicinity | Open virtual neighbourhood network to connect IoT infrastructures and smart objects. https://vicinity2020.eu/vicinity/
16. AWS IoT Core Features - Amazon Web Services. https://www.amazonaws.cn/en/iot-core/features/
17. Nakhuva, B., & Champaneria, T. (2015). Study of various internet of things platforms. *International Journal of Computer Science & Engineering Survey, 6*(6), 61–74.

18. Microsoft Azure Tutorial for Beginners: Learn in 1 Day. https://www.guru99.com/microsoft-azure-tutorial.html
19. Google Cloud Platform. https://www.conceptdraw.com/solution-park/computer-networks-google-cloud-platform
20. Stramowski, M. Philips Hue in Android. https://www.netguru.com/codestories/philips-hue-in-android
21. IoT Standards & Protocols Guide | 2019 Comparisons on Network, Wireless Comms, Security, Industrial. https://www.postscapes.com/internet-of-things-protocols/
22. Grant and Svetlana. (2016). *3gpp Low Power Wide Area Technologies*. GSMA White Paper.
23. Vinkovič, S., & Ojteršek, M. (2017). The Internet of things communication protocol for devices with low memory footprint. *International Journal of Ad Hoc and Ubiquitous Computing, 24*(4), 271–281.

Chapter 3
Business Models and Use Cases for the IoT

Carmen Perea Escribano, Natalia Theologou, Matjaž Likar,
Athanasios Tryferidis, and Dimitrios Tzovaras

1 Business Models

1.1 What is a Business Model?

Although the phrase *business model* is frequently used, its meaning is not always fully understood. There are plenty of definitions of business models available, from simple ones such as "All it really meant was how you planned to make money" [1] to more formal, for example: "a business model provides a holistic picture of how a company creates and captures value by defining the Who, the What, the How and the Why of a business" [2]. Moreover in [3], a business model is defined as a "representation of a firm's underlying core logic and strategic choices for creating and capturing value within the value network". Overall, a business model describes the strategy that an organisation creates to make money from its products or services. The internet era has brought new ways to obtain money and make a profit, as well as, mechanisms that were foreign before and we could not understand how to make a profit out of them.

C. P. Escribano
Research and Innovation, Atos Spain, Madrid, Spain
e-mail: carmen.perea@atos.net

N. Theologou · A. Tryferidis (✉) · D. Tzovaras
Centre for Research and Technology Hellas/Information Technologies Institute, Thessaloniki, Greece
e-mail: nataliath@iti.gr; thanasic@iti.gr; Dimitrios.Tzovaras@iti.gr

M. Likar
Gorenje gospodinjski aparati d.o.o., Velenje, Slovenia
e-mail: matjaz.likar@triera.net

© Springer Nature Switzerland AG 2021
C. Zivkovic et al. (eds.), *IoT Platforms, Use Cases, Privacy, and Business Models*,
https://doi.org/10.1007/978-3-030-45316-9_3

A business model reveals where the business is headed and the mechanisms that will get it there. In order for a company to write a good business model, deep thought and tough decisions are required. A business model is often one of the first tasks of a company, but it is also useful for established companies that are moving into a new market. A solid business model keeps product and company builders accountable for what they are working on and the time and resources they consume [4].

It is essential to carefully analyse the value that IoT is providing. In some cases, the value is provided by a new service, while in others the cost reduction is the one that brings significant value.

One of the main challenges that IoT companies face is to select the appropriate business model given the fact that IoT business models are evolving continuously. This section will give an insight into the business models widely used today.

1.2 The Impact of Internet of Things on Business Models

Together with IoT a wide range of new business models were revealed, which involve diverse partners of cross-industry ecosystems. For this reason, companies need to rethink their traditional strategies in order to stay ahead in IoT driven market environments. Business models, specific to a certain IoT ecosystem, can be seen as a means of communication between current and future ecosystems [5].

In the IoT context, the traditional elements used to describe business models need to be expanded with extended dimensions. In a dynamic and connected world, products have new features and functionalities that are constantly evolving together with the ability for new services, optimisation and customer service experience through over the air updates. A business model innovation becomes necessary as IoT-based value creation has now new possibilities [6].

Consequences that most traditional product-based companies are facing in the IoT era, according to [7], are in the domains of marketing (from transnational to relational marketing), sales (from selling physical products to selling service contracts and capability), and customers (from taking the ownership of physical products to purchasing services). In addition, the challenges with respect to business models and customer offerings include:

- Stakeholders perspective: understanding what value means to customers and consumers instead of only from producers and suppliers' view
- Services instead of products mindset
- Developing a service culture
- Business ecosystem thinking with multiple actors who interact with each other

1.3 Business Models Challenges

During the last years, many companies focused on embracing the IoT, however, these kinds of projects are not always successful. There are multiple challenges that should be tackled before or during IoT projects. These are:

- Absence of required knowledge: Organisations do not have employees with the required expertise to undertake IoT projects. In spite of new data availability acquired with the new systems, organisations are not able to exploit the value of the new data obtained to justify the cost of the investment made in the new project.
- Different architectures: In the end, each organisation is different; the set of different architectural components is very extensive and differs from one company to another. This leads to a situation where companies have different IoT projects and replication of IoT solutions is hard or even impossible.
- Interdepartmental collaboration: The IoT projects affect several departments in the company, from the operational to the business departments. Open collaboration and real understanding should exist between the different departments of the organisations which traditionally work as silos. To be successful in IoT projects a common understanding is needed.
- Administrative and regulation barriers: Administrative and regulation barriers are delaying the adoption of IoT solutions.

1.4 IoT Business Models

It is important to focus on business modelling and design methods that support traditional product-focus companies in their transition into the digital connected market. There are multiple ways of classifying IoT business models. In the following paragraphs, we provide a classification depending on where the business model is focused: on a sensor, on a value-added service (VAS), or on a platform.

1.4.1 Sensors Business Models

Some business models used on sensors are described below.

The Razor Blade This model was popularised by Gillette that sold the razors to a low price to create a continuous revenue stream from the selling of blades. This model was used in the IoT market. One example is *Amazon Dash Button*. And it is used nowadays by HP connected printers (https://www.hpsmart.com/us/en).

Pay per Use Usually, expensive products are released under this model. The sensors on the product allow the manufacturer to track the usage of the device. The Rolls Royce Power by the hour [8] is a well-known example of this business model. Rolls

Royce is in charge of the engine maintenance which is streamlined thanks to the sensors, airlines only have to worry to take passengers from a point to another point of the planet.

Data revenue The product with the sensors is sold to the customer, but the data collected by the sensors is also sold to other organisations or used in the organisation what sold it. An example of this business model is the Roomba plans to sell/share the house maps obtained by the vacuum in the houses [9].

1.4.2   Value-Added Services Business Models

VASs offer insight into the data generated by the Things. Typically, the VASs are provided to the customer through the following business models.

Customised Business Model In this case, the VAS is created ad-hoc for a customer and it is sold to the customer directly.

Package The VAS could be used in multiple customers and the revenue stream can come from a number of devices, messages sent using the service or through payment. The VAS is sold in a market place. The vendor obtains a percentage of the revenue obtained by the platform operator.

1.4.3   Platform Business Models

According to the taxonomy proposed in [10], revenue models for IoT Platforms have three dimensions with the following characteristics:

Pricing It is distinguished if the projects are charged or they are independent on whether the projects are developer or enterprise projects. For example, some prototype projects can be free and will be charged when the production phase is achieved.

Transaction-Based Revenues The following options are mentioned in this dimension "Per connected device", "Per API call", "Traffic based", "combination of multiple sources", "per request", or "free for use". This dimension reflects what parameters are used to measure the price.

Continuous Revenues It is examined if the revenue is obtained periodically (monthly, yearly) or depending on the growth of the implemented solution.

1.5   Data Driven Business Models

IoT is generating an immeasurable amount of data, however, organisations are not always taking full advantage of the information generated by these data. For example, data generated from a health wearable could be valuable for a health research company, as well as some relevant results could be obtained from them. Some companies are trying to obtain data from IoT devices and sensors [11], however, privacy concerns are arousing. With no doubts, insight obtained would be very valuable but the privacy of the individuals would be at risk. A balance should be found, in order not to miss the opportunity, while the privacy rights of citizens are respected.

Data marketplaces are an option to take advantage of the data. In these platforms, some organisations act as data providers, others act as data consumers and other organisations are in charge of the platform. These roles are not exclusive as the same organisation can provide data to be consumed and consumes data from other organisations. The organisation which is in charge of the marketplace should play a neutral role. Furthermore, VASs over the data can be commercialised by third parties, enriching the value of the data provided in the platform.

2   Internet of Things Driven Digital Transformation

2.1   The Impact of the Internet of Things on Society

The IoT adoption is being constant, however, the impact caused in the way of life is being breath-taking. Individuals, homes, cities, companies, and countries are incorporating new sensors soundlessly [12].

The IoT improves and will improve individual's life. From improvements in health monitoring and assistance to improvements at work (Augmented reality, safety, monitoring, etc.) and also improvements in the connected home [13].

Furthermore, it improves organisations productivity, saves costs (predictive maintenance, reducing response time, etc.), and transforms the way of working and cooperation since organisations cannot work in an isolated way anymore and collaboration between organisations is the only possible way not to be left behind [14]. Cities management is also improved in multiples ways (saving in consumption, agile maintenance, resources optimisation, etc.) [15].

So, multiple aspects of daily life are being impacted by technology. On the other hand, several cybersecurity attacks initiated in the sensors layer have exposed their vulnerability [16]. The relevance of some of the assets (Energy plants, health devices, water supply, baby cameras) "watched" by the sensors has triggered the alarms.

Therefore, it is a turning point where adoption and concerns should find alignment, and security awareness should be foster.

Another important worry is the waste produced from the sensors called e-waste, as sensors are incorporated into normal objects (watches, sports shoes, neck-laces), the life of the products is being shortened and more waste is being generated [17]. Furthermore, the energy-consuming of the sensors should be taken into account.

To finalise, the ethical concerns that have arisen around the use of IoT devices by humans such as privacy, security, and also safety continue evolving and society should be prepared to tackle them [18].

2.2 The Impact of Internet of Things on Digital Transformation

The digitalisation of society is a phenomenon we have been facing for the last decade and coincides with the development of technologies such as smart mobile telephony, mobile internet, IoT, cloud computing, data analytics, etc. All these technologies influence the evolution of the whole society, business entities, public institutions, and individuals. The impact on the economy is particularly noticeable, as the effects of digitalisation are visible in almost all industries. Digitisation has already taken place in the fields of communication, media, entertainment, tourism, advertising, commerce, education, publishing, film and photography, and many other areas where digital opportunities for business improvement have been recognised [19].

In many mature industries, such as consumer electronics, it is very difficult to retain sustainable revenue over time using just traditional product development approaches. Even the best industrial design, new components, or new innovative features cannot be effectively protected from the intellectual property rights perspective. Competitors can usually provide similar solutions in a very short time after the initial market launch by pioneers, so the original provider can harvest the benefits only for a relatively short period. For that reason, companies are looking for possibilities on how to define new business models, utilise new technologies, and develop more complete sets of solutions to create added value for the customers and to generate more stable revenue.

IoT is a technology that has been widely adopted. It has been used as a technological enabler for digital business transformation. Because of the opportunities brought by interconnected things and interacting with people, shopping is also changing. Previously, buying was a one-off event. The customer was largely unknown to the manufacturer or to the seller after the purchase. With the introduction of connected products, the purchase and use of a product or service is a continuous "non-stop customer" business relationship. Let us look closer what are the IoT enabled drivers for the digital business transformation.

Marketing IoT provides customers insight, enables understanding of customers' behaviour, improves customers' loyalty, supports better-targeted marketing campaigns, digital marketing, and social marketing.

Sales The traditional manufacturing industry is shifting towards the services industry. IoT enables new business models and as a result new sources of revenue. Businesses can monitor sales vs. geography with very frequent updates.

In this interconnected ecosystem, companies can also partner with other verticals (like retailers, insurance companies, energy suppliers, home automation providers, surveillance services, etc.).

Aftersales With IoT, companies can more precisely understand customer behaviour (How they use products? When a replacement purchase can be expected?). Companies also have better insight on how the products are performing in a real-life environment (e.g., efficiency, utilisation, number of cycles, durability, diagnostics, preventive/predictive maintenance, etc.). There are also plenty of new sales opportunities such as offering spare parts, accessories, or supplies.

R&D Insight into a big sample of real-life data (understanding what works and what does not; what product's functions users really need, etc.) can be used pragmatically in the new product innovations.

Artificial Intelligence, Data Analytics, and Machine Learning These disciplines and techniques require a new set of knowledge and skills for the organisation. In combination with technologies like IoT, these techniques represent an important asset for companies that are better prepared for new business challenges.

3 Cross-Domain Use Cases and Value-Added Services

In this section, some of the current challenges and problems that countries worldwide are facing nowadays in domains such as building, transport, energy, health are analysed as well as use cases and VASs for each area, which facilitate in solving these issues. The implementation of the use cases highlights how advanced techniques can facilitate the creation of diverse services across IoT domains.

But What is Actually VAS? VAS could be defined as a piece of software that implements an algorithm from a simple calculation/data processing to some advanced techniques such as clustering/big data analytics, data storage, and auditing. VASs collect data, in order to further process them, from the IoT infrastructure (IoT devices, sensors, etc.) and are based on the available IoT data from other IoT infrastructures. They further reveal a business model potential/commercial exploitation of such a service (e.g., for application developers, service operators).

3.1 Smart Buildings

3.1.1 Motivation and Challenges

It is critical to improve resource management, resource consumption, and predictive operations in buildings and organisations operating and managing commercial buildings. Utilising wireless technology is relevant both in the case of old and new facilities. Battery-powered units are especially helpful to ensure quick and cheap installations that are independent of existing electric infrastructure. Older facilities, lacking in technological upgrades and modern information technology infrastructure, can benefit from light-weight, wireless IoT devices to provide improvements to energy efficiency, resource management and to gather inputs for daily operations.

As of 2017, 35% of buildings in the EU are more than 50 years old, and their energy efficiency is significantly worse compared to new ones according to [20]. Countries with the largest components of older buildings include the UK, Denmark, Sweden, France, Czech Republic, and Bulgaria [21]. Moreover, they often lack modern facility management tools, such as intelligent heating, ventilation and air conditioning (HVAC) control and monitoring, as well as another industrial control system (ICS) functionalities. Buildings are responsible for 36% in terms of CO_2 emissions and for 40% of total energy use [21], while they use 40% of global raw materials [22]. However, barriers towards more efficient building infrastructures for sustainable construction exist and are split incentives, lack of consistent, accurate data analysis and interpretation, lack of knowledge about sustainable construction, high upfront costs, and regulatory gaps, as identified in [22].

For modern buildings, IoT solutions will pave the way for solutions to come. Large scale deployment of smart meters in buildings is an important step for achieving energy efficiency and management. IoT-enabled networks and a flexible infrastructure could be capable of integrating dynamic control strategies, providing a significant opportunity for energy savings in the building domain [23]. Also, cutting-edge virtualisation and big data solutions are utilised to implement efficiency and allow dynamic extension of the system with new sensors. Governments and regulatory agencies around the world are taking actions towards the energy efficiency of commercial buildings, with leveraging IoT technologies enabling new business opportunities and revealing benefits to all relevant stakeholders. More specifically these technologies are expected to enhance the functionality, capabilities, energy efficiency, and cost-effectiveness of buildings [24].

3.1.2 Use Cases

Predictive Operations Keeping buildings clean and ready to use requires significant human resources. The work is based on predefined schedules, and cleaning teams spend much of their time on routine rounds to check if rooms and toilets need cleaning. IoT solutions hold potential to give input to employees and subcontractors to make them more efficient, for example through reducing the amount of time they spend on rounds by redirecting them to where they are most needed. Since cleaning and waste management crews usually have no reliable source of information on how much a room has been used before they visit it, cleaning and waste removal services are time-based, with a schedule that states how often a room should be checked and the last time it was visited by the crew.

The time interval is an imperfect proxy for the need for cleaning and waste removal: How dirty a room is or how much waste has been generated does not primarily depend on how much time has passed, but on how many times the facilities have been used. In the absence of usage statistics, the crew must regularly and manually check the status of all the rooms in a facility to make sure that the rooms and toilets are clean and that the wastebaskets are empty. Moreover, the crew has limited input as to which parts of the facility have been used most, and it is, therefore, challenging to target their efforts towards the areas that need their attention. By introducing the possibility for semantic interoperability such as the VICINITY platform, different IoT platforms can provide data to be used for software development in an easier and real-time environment.

The VAS Predictive operations developed for integration to interoperability IoT platforms, such as VICINITY, provide the cleaning and waste collection teams with information and alarms that can improve their efficiency and the quality of their service offerings. The personnel are able to access the information and their settings through an online user interface. Through the interface, they can view real-time and aggregated data, as well as graphical and statistical analysis. The team can set and customise alarms with threshold values for the number of person movements. An alarm is then issued if a toilet or room has been used a certain amount of times. In this way, the personnel receive a notification that it is time to clean it and/or pick up the garbage. In summary, the service helps to reorganise cleaning and waste removal from a frequency-based to an on-demand system.

Resource Management In addition to human resources, the management and operation of a building require significant amounts of consumable inputs, such as energy and water. IoT solutions hold the potential to conserve and redistribute these resources while accommodating tenants' needs and expectations. This makes data collection easier for the management team, and helps them exploit considerable cost-saving potential in their electricity consumption. Moreover, they are able to accurately bill their tenants and discover water leaks or other abnormal consumption patterns early. The service supports their daily operations and decision-making on energy management, load optimisation and heating, HVAC control by distinguishing typical from non-typical situations. Combined with weather forecast data, the service can predict upcoming electricity peak loads based on estimated needs for additional heating or cooling of the premises.

The service can provide considerable value to managers by helping them harmonise electricity loads to save on load factor tariffs, as well as to discover statistical trends on how consumption changes throughout seasons, weekdays, and times of day for improved operations planning. If an upcoming peak load is identified, the managers may shift or shed loads immediately, e.g., by temporarily turning off or down heating, cooling or ventilation systems. Moreover, automatic consumption data collection gives value to managers by replacing time-consuming and sporadic manual meter readings and enables them to bill tenants precisely and transparently.

As such, the service brings four key values to building managers:

1. Collection of consumption data for billing and statistical overview, as well as input to operational scheduling of resource-consuming systems; providing value through time-savings, tenant billing, and improved future operations.
2. Providing alarms about non-typical situations, so that periods with, e.g., unusually high electricity load or abnormal water consumption can be discovered and abated.
3. Providing value by enabling electricity load harmonisation for reduced load factor tariffs and early warnings about potential water leaks or tenant negligence.
4. Providing automatic control signals based on non-typical situations such as alarms, and automatic control signals that enable harmonisation of electricity loads.

3.1.3   Use Case Business Modelling

Predictive Operations Statistical usage trends combined with alarms make it possible to target cleaning according to use and identify critical situations. Team members can be notified directly if a certain usage threshold is exceeded and take immediate action. This action reduces the need for systematic checking of rooms and allows the team to concentrate their efforts on where they are most needed, thereby saving time and helping maintain a high-quality service. Moreover, cleaning and waste removal can be limited or postponed in areas with little or no use.

This use case arouses possibilities in selling and partnering with other actors such as building managers, building owners, utilities, building tenants and an income from Software as a Service (SaaS), licences, SLA, upgrades, and transaction cost. In building management and in old buildings, one of the assets of this technology is that the equipment/sensors are battery-powered and therefore independent of an electricity supply, which means it can be installed everywhere.

Resource Management This VAS makes data collection easier for the building management team, and helps them exploit considerable cost-saving potential in their electricity consumption. Moreover, they are able to accurately bill their tenants and discover water leaks or other abnormal consumption patterns early. Although the VAS service only consumes data from water meters, electricity meters, and electricity meter readers in the demonstration case, the range of data sources could be expanded in a commercial version.

The VAS for energy resource management could be commercialised depending on the feedback and result from the deployment. This VAS can be bundled together with equipment for measuring energy and water consumption. Possible customers for this solution could be building managers, building owners, utilities, and building tenants. The product to be delivered to market would be a SaaS with licences, SLA, upgrades, and transaction cost.

Moreover, real estate owners have a great involvement in smart cities and smart buildings and have a primary business goal to promote their activities and results in this domain. Small companies delivering services to smart buildings like maintenance, catering, parking service, and others see a possibility to give better services and be more efficiency by being able to have up to date information of the usage of the building, office rooms, institutions, and parking spaces. Rationale of the use cases could be concluded in the following:

- Strategy to be innovative in smart buildings
- Focus on energy efficiency
- Focus on energy costs
- Focus on leaks and overload of energy usage
- Focus on optimisation of buildings usage
- Focusing on better service to tenants/building owners
- Saving man hours for maintenance and cleaning

3.2 Smart Transportation

3.2.1 Motivation and Challenges

Parking space is an example of a resource that might be in high demand for certain periods. Vehicle drivers that are out for buying grocery stores, people that are paying relatives a visit, and tourists that are sightseeing want to find an available parking space. In some cases, it may even be a matter of life or death as when blue light agencies (ambulance, police, firefighters) have received an alarm and have access to a pre-reserved parking spot.

In the report "Spaced Out: Perspectives on parking policy" written by John Bates and David Leibling [25], it was stated that the average car is parked at home for 80% of the time, parked elsewhere for 16% of the time and is only on the move for 4% of the time. These findings were based on English figures, but it is safe to assume that it holds true for most other countries as well. An unused parking space does not only represent dead capital to the owner of the parking space but represents also a loss to the community since more people will stress and put a challenge to the infrastructure by drivers trying to locate parking space in other places.

Cities are suffering from congestion emission and poor air quality leading to reduced quality of life. Society loses income through poor resource management of unused parking spaces which are mostly owned and administrated by the building cooperatives and are only made accessible to tenants. Unused parking spaces could be valuable to others with the possibility of finding suitable parking spaces with certain properties (dimension, EV, handicap, etc.) as people are struggling to find proper, safe, and secure parking areas in Europe without the proper technological tools in hand, like mobile applications, to facilitate them.

New technologies are being used in order to solve the problem of driving around looking for a parking space. Transportation problems are one of the major issues that leads to the creation of smart cities as analysed in [26]. Open research issues for smart parking are congestion management, outdoor parking management, driver guidance, etc. Steps are being implemented worldwide in order to resolve these challenges by deploying smart parking management systems in order to improve the operational efficiency of cities while optimising time, cost, and reliability [27]. As described in [28] projects that have already developed and deployed in cities like in Nice, still have open issues to handle. Some of them are the robustness of sensor devices, the stability and timeliness of sensor networks, proper management of resources, information dissemination, and uncertainty factors.

3.2.2 Use Cases

Shared/Priority Parking This use case has a specific focus on managed healthcare apartments, and demonstrates how to transport information and building data can be integrated with assisted living through agreements with car space owners and other stakeholders. In the demonstrated solution, prioritised parking space, booking, traffic analysis and messaging services based on authorisation and access data are managed according to conditional rulesets. The use case presents a service that allows neighbourhood residents to share their unused parking spaces for shorter or longer periods. All parking spaces in underground garage facilities are owned and administrated by the buildings cooperative. The available parking space is only made accessible to tenants. However, in case they have accepted being part of a rental agreement for parking space, they are free to assign or reassign the use of the parking space as they like as long as it does not go against the regulations. When accessing the garage facility, it will always be the tenants that have priority.

The smart parking sensors report proximity and temperature. These are included with datasets from stemming from external traffic sensors and open data. The datasets are used for big data analysis to identify usage profiles and generate traffic forecasts that are available to building managers, traffic controls, and virtual neighbourhoods. Booking, configuration, and status of parking space are handled through different devices and user platforms such as mobile apps. As such, the statistics are used for building management resource planning from building managers, residents, and parking space owners. They also shape the basis for improved services to vehicle owners, visitors, and agencies being private or public by nature.

eHealth Emergency Parking This is another use case that also demonstrates how basic functionality for shared parking space can offer new business models and exploitation potential when using interoperable ecosystems and VASs for mobility and assisted living. In building clusters, tenants with certain demands due to age or disabilities may need special assistance. In order to make work of health care assistants that may take care of situations that might arise most efficiently, they need parking space close to their clients.

The tenants have health care equipment from different vendors and the use of the interoperability platform is of crucial importance. This use case is an extension of the previous one, where priority parking is assigned to healthcare personnel according to medical or other significant alerts received from IoT assets installed at buildings.

3.2.3 Use Case Business Modelling

The VASs provide real-time information about the situation at the parking space and status messages from apartments with smart appliances. This allows for assigning parking space locations based on priorities and location, and activating events like locking down parking space orders in case of emergencies. The goal is to provide the best possible parking space based on properties, preferences, and priorities in order to ensure a both environmental and economic viable alternative to driving around the block looking for vacant parking spaces and perhaps also receive fine tickets during this process.

Thus, customers encompass both public and private representatives, ranging from homeowners to municipalities and counties. In general terms, all that own a parking space or a parking lot, and all that are looking for a parking space close to an intended destination can be defined as a customer. In the previous use cases, these customers have been narrowed down to residents that are owners of parking space and medical staff on-site as well as visiting caretakers. Secondary customers may be value providers and developers of smart appliances and sensors that are related to parking. By bundling several solutions it opens for access to new markets. These markets will require consultancy, maintenance, upgrades, scalability, hosting, and integration on different levels. Different kinds of software along with mobile apps and administration systems will also be part of the offer for customers.

The business model identified is a subscription-based model that is tailored for larger facilities and operators who manage a larger set of areas suitable for parking. Not all areas need to have a sensor installed. The subscription-based model offers instead a scalable solution where new tools and services can be added as the product evolves and customer demands change. Such tools would be statistics, ownership models, scheduling tools, etc. It will typically be users from the municipality and other larger organisation that will benefit from this offer.

The services offer functionalities that support different emergency levels. This includes locking parking spaces for emergency parking and reassigning vehicles that already have booked, but are still in transit to another parking place. Integration of smart light that changes colour and frequency of blinking based on arrival, remaining time, and errors is also implemented and supported for further extending VASs. Integration of a parking app offers even more functionalities—including recommended parking nearby, available space, specifying favourite destination and also supports transactions. This means that to each parking space a value can be assigned and source of income could be offered to the owners of the parking space. The parking app also differentiates between parking space and parking lots—which do not necessarily have to be situated at the same physical location. The parking app can support different user roles (owner, driver, medical) and different vehicles (cars, vans, motorcycles, EV, ambulance, bike) which has different specifications and expectations for the parking space and surrounding areas.

Finally, the system can offer usage statistics and transaction lists sorted on owner, driver, vehicle, parking space, and parking lot. These data can be used to further identify where and when incidents can be expected to occur more frequently than other places. The statistics will thus serve as a good tool for planning and allocating resources—both when it comes to caretakers and when it comes to parking space.

3.3 Smart Energy

3.3.1 Motivation and Challenges

Energy efficiency especially in public buildings is a major issue and improving energy performance is a key action in the fight against climate change. Barriers, as identified in [29], are lack of knowledge of citizens and building occupants on how to conserve energy, lack of problem awareness, and lack of incentives. In the meantime, the cost of fossil fuel usage fossil is higher than before and its negative impacts on the planet's climate and ecological balance make it important to explore new clean-energy sources and improve the energy efficiency in the consumer-side smart grids of various buildings [30]. However, a smart network of sensors can provide a tool of solutions for energy efficiency by sharing data and information improving network management. As described in [31], generating reports and performing alarms based on data analysis is a need for both consumers and suppliers in order to raise awareness.

The concern with the health and well-being of the community has been an ever-increasing priority, and this care is already directed towards the improvement of the indicators that guarantee the quality of buildings. This trend is supported by public and private entities, raising the importance of Indoor Environment Quality (IEQ) scrutiny, which interconnects the reality of the construction and (re)adaptation of more energy-efficient buildings.

IoT technologies could facilitate cross-domain use cases such as the combination of energy domain and health cases. There are several documented benefits from ultraviolet solar radiation (solar UVR), however, a lot of adverse effects are described, including the occurrence of photo-induced skin cancer and negative effects on the immune system. By leveraging IoT sensors and devices such as wearable or smartphones with related applications, which integrate satellite-based georeferenced radiometric data with meteorological and other useful data, it is possible to alert users of critical situations as described in [32].

3.3.2 Use Cases

Energy Efficiency and IEQ Management in Municipal Cluster of Buildings
Buildings in general and municipal buildings, in particular, are used within the originally designed and designated parameters without any consideration of space optimisation that can be achieved by taking into account the measurement of the consumption of resources by each space and other parameters, such as IEQ which is based on parameters such as light, CO_2 accumulation, or space use optimisation by users. This link between the space use, IEQ quality measurement, and resources consumed by buildings optimisation is at the core of the use case.

In this use case, multiple types of sensors and equipment are used to collect information from the buildings in real-time. Parameters such as temperature, CO_2, humidity, noise, motion, and energy consumption are considered for further processes modification and analysis, some processes automation may be considered and analysed for viability. These parameters are also combined with data from a weather station. The primary goal is to facilitate dynamic data collection from various sources, including sensors and other cloud services with the goal of creating a new service. In Fig. 3.1 a narration of IEQ services for smart schools is presented. The service aims at creating a new dynamic platform approach of monitoring and analysis of building performance, resources and use and systems performance evaluation. Smart School service aims at delivering IEQ services where CO_2 level in the classrooms could be managed by teachers and hence directly improving study conditions.

The collection of data becomes useful for various decision-making processes, such as equipment substitution and upgrades of municipal assets/equipment, opti-

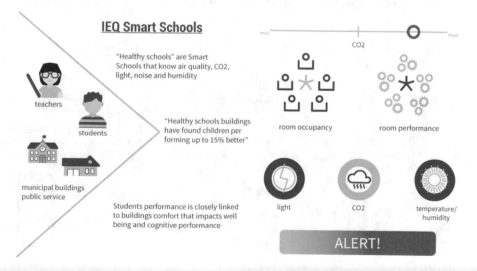

Fig. 3.1 IEQ services for smart schools

mal space utilisation and functional use, energy and other resource management, daily consumption optimisation, including peak shaving and subsequent integration into energy flexibility systems. Well-functioning demand response models and eventually fully integrated smart grids would depend on well-integrated building blocks, which start at the building level, scaling up to the buildings cluster level and further on onto neighbourhood, city, and region. Use case aims at the municipal cluster buildings level with the goal of scaling up.

Ultraviolet Radiation (UV) Info Services for Citizens and Tourists Services are designed and developed to meet the need for information services provided by municipalities within smart cities or towns context for health benefits and local conditions awareness towards citizens and tourists. The goal is to create a way of leveraging the existing equipment within the region, town, city that can provide data for the creation of local IoT enabled VAS. These VASs that are built on the unleashed data could be processed and made available to tourists and citizens geographically located in the area. By creating these services, it is possible to improve current indicators and augment simple weather conditions with advisory and more relevant services.

In addition to the services described, additional cases can be proposed, such as leveraging existing Closed Circuit TV (CCTV) security cameras for wildfire identification using computer vision and machine learning techniques.

Narration of the use case is presented in Fig. 3.2.

As an example of the usefulness of this kind of service, in very sunny regions attracting tourists from Europe and beyond, it is increasingly common to find tourists with (severe) sunburns. By using this kind of informational service throughout the region, it can create awareness and make the vacation experience much safer and more comfortable for all tourists while reducing the number of persons in need

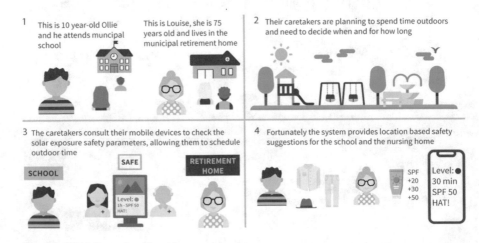

Fig. 3.2 UV info services for citizens and tourists

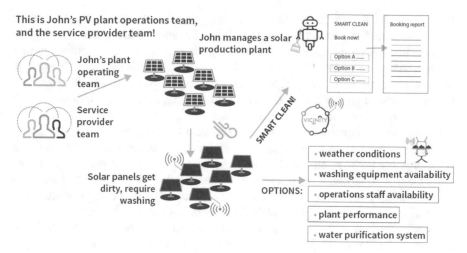

Fig. 3.3 Smart clean for distributed renewable assets

of hospital care due to sunburns. The service can also be utilised in retirement homes to decrease the exposure of vulnerable citizens and in public schools.

Distributed Energy Assets Management Operations and maintenance of distributed energy assets could be optimised through the deployment of IoT enabled automated services. This constitutes a large portion of O&M of systems expense. Further penetration of distributed energy resource (DER) provides further opportunities for scale and optimisation. When the owner or operator of a PV plant has access to information such as weather conditions, solar radiation, and PV production, they can more easily identify problems with the PV plant, and improve optimisation including better scheduling of systems maintenance.

This use case intends to optimise resource deployment and the resulting scheduling of cleaning of the panels by indicating the best possible dates for panel cleaning taking into account several different parameters such as the high concentration of soiling on the PV modules, predictability of rain, solar resources, human resources availability, and the related equipment readiness. The service has the potential to be rolled out inter-regionally for distributed renewable energy production facilities: Automated cleaning service "Smart Clean" for distributed renewable assets in operation. A narration of the use case is presented in Fig. 3.3.

3.3.3   Use Case Business Modelling

Energy and resources management VASs were approached from various angles. On the one hand, testing equipment data share valorisation for local use, to increasing value in neighbourhood clustering of municipal buildings for resources consumption optimisation to designing and development IEQ based services for public schools.

Specific emphasis is made on wider users' engagement in understanding the interior environment and impact on cognitive abilities. Leveraging this experience in IoT systems and operation expertise in managing renewable energy resources generation capacity, an additional service was envisioned for digitalisation operations and maintenance, specifically focused on cleaning of concentrated photovoltaics modules with the goal of rolling the service to other parks and installations.

A potential for co-development of applications via an enabled marketplace with leading precision equipment manufacture can lead to opportunities beyond current VASs. Implementations of such metering and analysis systems are possible today but everything has to be custom designed which does not result in scalable solutions since there is no incentive to make devices interoperable. A company that wants to create such services can do it but there is no incentive for other companies to reach a common communication protocol for interoperability of devices. Infrastructure as a Service (IaaS) will allow a shorter time-to-market as service providers will not depend on data from a single manufacturer since any standardised data flow can be replaced by another of the same nature, even if the equipment is made by a different manufacturer and communicates in a different native protocol.

An analogy between an IoT interoperable platform such as VICINITY and current solutions can be made by comparing Sigfox[1] and LoRa.[2] Sigfox has all the communication infrastructure ready and one can just set up a device and use the network, paying for data but getting a much shorter time-to-market. On the other hand, with LoRa you can invest in implementing your own network infrastructure and not pay anything for data, although the time-to-market is much longer.

The public service could be assessed for scaling up and could serve as a platform for additional complementary services provision. IaaS and regional Labs for public services rollout would be proposed for consideration. The total cost of ownership of the system would depend on the cost of systems set up, components, cost of connecting to an IoT platform and running costs.

3.4 Smart eHealth

3.4.1 Motivation and Challenges

Healthcare is becoming more and more difficult to manage due to insufficient and less effective healthcare services to meet the increasing demands of the rising ageing population with chronic diseases. As analysed in [33] today's challenges for the health domain are summarised in the following:

- Limited quality time for checking patients
- Adherence monitoring of prescribed treatment

---

[1] Sigfox—https://www.sigfox.com/en.
[2] LoRa Alliance—https://lora-alliance.org/.

- Increasing ageing population
- Urbanization
- High demand for healthcare workers
- Rising medical cost

For this reason, IoT applications in the eHealth domain are increasing. Healthcare applications are moving from hospital centred applications towards patient-centred applications.

It is vital for governments to promote disease prevention, better and effective management, and rehabilitation of older people. In 2014, Europe had a 31.4% share of elderly people who live alone [34]. The cost of residential care can vary considerably by location and country and depends on whether someone requires nursing or dementia care. The solution of a nursing home is necessary in some cases due to everyday conditions and obligations or serious illnesses. When relatives of the elderly work long hours, their obligations often keep them away from home and it is very difficult to be able to properly take care of them. It is also a necessary solution in cases of health issues that need continuous medical care, such as when suffering from Alzheimer's disease or a severe disability. However, it is preferable to keep the elderly relative at home, assisted by medical personnel. Assistance of home-based elder support systems is important and a useful asset for doctors and relatives to judge the health status of the elderly and to provide the necessary assistance. This could lead to the avoidance of long-term unwanted events and diseases.

Moreover, Europe has a high proportion of overweight people, 51.6% according to [35], so EU actions are focused towards promoting better nutrition and physical activity in the EU [36, 37]. Unfortunately, today's way of life includes sedentary jobs and a lot of hours in front of the computer and/or television.

Life satisfaction (a measure of subjective well-being) is lower in Greece than the average for the European Union according to the World Health Organisation (WHO) [38]. According to the same source among objective well-being measures, 61% of people aged over 50 years reported that they had relatives or friends on whom they could count when in trouble, which is among the lowest proportion in the region. A national health policy aligned with Health 2020, including an implementation plan and accountability mechanism, has been developed but not yet formally adopted. Moreover, WHO estimates that in 2014, 66% of men and 55% of women in Greece were overweight (body mass index ≥ 25), with an increase of 2% points since 2010 for both sexes. In addition, the prevalence of obesity (body mass index of ≥ 30) increased by 2% points. The averages for the region are similar to that in Greece regarding the share of overweight women (55%) and the share of obesity for both men (21%) and women (25%) according to [38].

Use cases described below are adopted in the municipality of Pylaia-Chortiatis. The municipality has 70.110 inhabitants (2011 census) [39] in an area of $167.800 \, km^2$ and is a pioneer in eHealth services in Greece. It is a member of the "National Inter-municipal Network Healthy City—Health Promotion (NINHC-HP)", which is certificated by the WHO as a National Range Cities Network and it is also offering a program for elderly and middle-aged people called "e-Help at Home", in order to allow Pylaia-Chortiatis citizens to live an independent and healthy life. The program offers monitoring of the weight and blood pressure of citizens registered to the program, as well as notification services in case of detection of abnormal events (panic/fall button, positioning of an Alzheimer patient).

3.4.2 Use Cases

eHealth and Assisted Living for Elderly People at Home The municipality deployed homes of elderly people living alone to remotely monitor their routine tasks, everyday activities, and medical data in order to implement assisted living for them and feel safe at home. The VASs related to the use case create applications for health care providers/municipality to track, monitor, and cluster data of people participating in assisted living programs while privately preserving them. These data are collected from IoT devices and sensors integrated into houses. Furthermore, a VAS that will be implemented for this use case is to detect abnormal detection at elderly people's homes. Therefore, in this use case building sensors are combined to leverage assisted living and eHealth at home with real-time analysis of information collected from building IoT sensors at home (motion detectors, smart appliances, etc.) which leads to a cross-domain use case.

Specifically for this use case, three VASs are implemented in order for the data from sensors to be processed and the various actors to have access to the data or services they are concerned with. The first VAS gathers the data coming from houses, meaning the data from building and health devices deployed in-house in all cases and privately stores them in order to be available for further processing by other VASs. Further scope of this service is to strengthen and unify data protection and processing of personal data for all individuals within the use case complying with the GDPR. Auditing of the data transaction, which is implemented, is essential for privacy issues. Specifically, the service includes auditing of the data transaction and access control mechanisms for the user (e.g., a relative of the elderly citizens) to have absolute control of who can access the data (the concept of consent). In addition to the latter, data access mechanisms are developed introducing the concept of consent, meaning that a user is able to control and manage who has access to his/her data. The GDPR "right to be forgotten" is of essential importance for the user and so relevant functionality is developed that will delete all audit logs and data related to a user that desires to abandon the project and delete his/her account.

Health care providers have the potential to process their patients' data through a user interface, exploiting the second service, which provides clustering of the available medical data. The goal of this service is twofold. It collects medical data of elderly people living alone allowing health care providers and relatives to know their current condition in order to provide advice if needed and check whether there is aberration. The service also handles historical data of medical devices and sensors deployed in-house in order to create behaviour profiles and help in clustering elderly people to different medical groups. This VAS is also an asset for relatives of the elderly to access their historical and real-time data and be aware of their condition.

Finally, the goal of this third service is to leverage IoT sensors, in order to provide a home-based solution that is able to detect behavioural changes of elderly people who live alone and offer an independent life. It relies on data coming from integrated building IoT sensors, which are processed, in order to produce context-aware metrics that are used for creating behavioural models per individual. This asset can detect significant changes in the elder's behaviour at home and provide information concerning the time and place where the abnormal behaviour was observed, in a daily basis, e.g., the elder starts spending more time in the bedroom or becomes more active during night hours than usual, which may indicate the beginning of health related or psychological problems. Any abnormal behaviour detection will trigger a notification to the mobile application of the health professional and/or relative of the elder.

Health Improvement for the Middle-Aged People   The focus of this use case is to promote a healthy lifestyle to the middle-aged citizens through gamification system. Its goal is to help the citizens to adopt new healthy habits and thus preventing, as much as possible, future health problems, meaning fewer visits to health care providers or dietitians and less primary institutional costs for health services. The above, require that middle-aged citizens of the municipality should change their everyday habits, participate more in athletic activities and compete for their performance with other citizens in a so-called "urban marathon".

Specialised staff (pathologist, dietitian) monitor the middle-aged citizens' blood pressure and weight measurements as well as their exercise data on daily basis and examine their improvement. For this purpose, equipment such as wearable activity trackers (fitness trackers) and weight and pressure monitoring devices are delivered to the citizens that participate in this use case, and beacons are placed at the municipality's athletic facilities in order to monitor data concerning the participants' visits. Moreover, citizens took part in a municipal-scale competition ("urban marathon") to compete with other citizens on health improvement achievements (e.g., citizen A ranks at the top 10% of citizens, according to the miles he/she has walked this week). These means are used in order to motivate citizens to participate even more in this health improvement use case. Citizens are informed of the status of their activities by using a mobile application which will be also used in order to sign up for this urban marathon.

Finally, yet importantly, the municipality has access to anonymised statistical data concerning its middle-aged citizens' health status and improvement over the period of this use case. Specifically for this use case, three VASs are implemented to overcome the challenges described in previous paragraphs. First VAS is responsible for gathering the data coming from health devices/sensors deployed in sports centres or worn as accessories, and further storing them in order to be available for further processing by other VAS. A further scope is to strengthen and unify data protection and processing of personal data for all individuals within the use case complying with the GDPR, similar to the first VAS of the previous use case.

Second VAS provides individual statistics for each participant according to the data that are acquired from the first VAS, such as weekly or monthly statistics for the citizen's improvements. Moreover, it calculates the participant's rank among the other participants, according to general statistics that are provided by processing citizens' data. Third VAS is responsible for providing an aggregated statistical analysis of anonymous citizen data concerning their health status and improvement.

3.4.3 Use Case Business Modelling

eHealth and Assisted Living for Elderly People at Home This pilot case is important for European citizens, as it will set an example for other municipalities, hospitals and institutional homes to follow. Use cases proposed could assist in elder people living in Europe and generally in the world.

It is vital for governments to promote disease prevention, better and effective management and rehabilitation of older people. Use cases regarding eHealth and assisted living can assist in this problem and facilitate with expenditures of hospitals, health institutes and health care providers. Value propositions of use case eHealth & Assisted Living is that participants in assisted living programs have the ability to remotely monitor their health by specialised staff and stay at home instead of caring institutions reducing primary costs for citizens and municipalities that have implemented the use case. Municipalities, call centres, and private health institutes will be offered with services regarding abnormal behavioural detection services in order to prevent unpleasant situations for elderly people.

Regarding the evaluation of the business opportunities offered by the respective use cases demonstrated in the eHealth domain in the Greek pilot case, there seems to be a prominent case in improving elderly people's life both in Greece and in Europe in general. In particular, in Greece, it is of key importance to seek for new alternative ways in order to provide new solutions in the health domain, given the low access to quality health services to elderly people. In the last couple of years, it has been made apparent by the Greek Ministry of Health that there is changing policy in Greece towards promoting eHealth services, in order to improve healthy life expectancy while dealing with diseases and tight budgets. On a larger scale, use cases regarding eHealth and assisted living services can contribute towards this problem and facilitate with expenditures of hospitals, health institutes, and health care providers. Value propositions of such use cases of eHealth & Assisted Living at home lies on the fact that participants can have the ability to remotely monitor their health by specialised medical staff while staying at home, instead of needing to move in caring institutions, therefore leading to reduced primary costs for citizens and municipalities.

The VASs derived from the use case could be distributed as a SaaS distribution model for the IoT platforms and offer data analytics service on top of collected data to municipality. The outcomes of the solution are expected to be presented to the regional government with the objective to be considered for further replication and further development in other municipalities.

Health Improvement for the Middle-Aged People As it was described in the introduction, the need for such use case is big, since obesity is increasing more and more each year, causing numerous side effects and health problems. In this user case, the municipality takes action by providing to its citizens a "service" for improving their healthy life and prevents future health problems. Citizens benefit from this service not only in their future body and health condition but also in the earnings that they will have by needing less medicine or fewer visits to health care providers or dieticians. This service is valuable not only to the citizens themselves but also to the municipality as it aims to reduce future health service costs.

On a larger scale that breaks the municipality boundaries, similar competitions could be organised on a national level, or any kind of larger scale, so that municipalities compete with each other. In such way, except the benefit to offer a healthier lifestyle for their citizens, municipalities could also have various predefined benefits themselves as well as reduced long-term costs in terms of public health services, etc. thanks to the improved condition of their citizens.

In eHealth, gamification solutions perform as an engagement for following a treatment plan and a motivation to stay committed by earning points with the intention of achieving a behaviour change in citizens' lifestyle. These solutions have gained considerable interest as access to healthcare resources has increased. Moreover, built-in measurement systems in IoT devices make it easier to access and sync real-time data for further processing and visualisation in corresponding applications. Pointing and rewarding systems boost the intention of people to include exercise in everyday life. Promising applications utilising gamification in the eHealth domain give great value and business opportunity to this use case.

The VASs derived from the use cases could be distributed as a SaaS distribution model and offer data analytics service on top of collected data to the municipality. These services could solve issues of affordability, accessibility, reliability, and technical support. The aim is to stimulate preventive actions smartphone/wearables for exercise.

4   Description of the Business Models Proposed by VICINITY

The VICINITY project has explored a wide range of business models. There is not a unique way to provide value to our customers and the different models could be applied depending on the customer needs. In this section, the description of the Business Models developed is provided.

4.1   Platform as a Service (PaaS)

This is the core service offered by the VICINITY platform. In this case, the
VICINITY Platform is provided as a service to the customer. The objective of
the VICINITY architecture is to facilitate interoperability between different IoT
infrastructures devices and to software-enabled VASs through a peer-to-peer (P2P)
network of VICINITY Nodes as presented in Fig. 3.4.

Each VICINITY Node provides access to IoT infrastructure and/or VAS. Once
the IoT infrastructure is integrated into the VICINITY Neighbourhood through the
VICINITY Node, devices connected to the infrastructure become accessible through
the VICINITY Neighbourhood Manager in the VICINITY Cloud. In VICINITY
Neighbourhood Manager IoT infrastructure owner can define sharing access rules
of devices and service (i.e., has direct control over his or her devices). Based on
these rules he or she creates a social network of devices and services called "virtual
neighbourhood". Customers are provided with plans that are based on the number of
messages (and not on devices). This gives them an opportunity to reach significant
savings comparing with the generic customers that are expected to pay per device
prices. Moreover, VICINITY is offered in the form of SLA for PaaS (Platform as
a Service) under certain limits of consumption. The PaaS is a technical service that
is based on an SLA with measurable technical parameters where partners can raise
tickets. The customer will access to their sensor data and also to others' sensors data
if the data is shared with them. VICINITY provides a great opportunity to obtain
insight of information that was not available before.

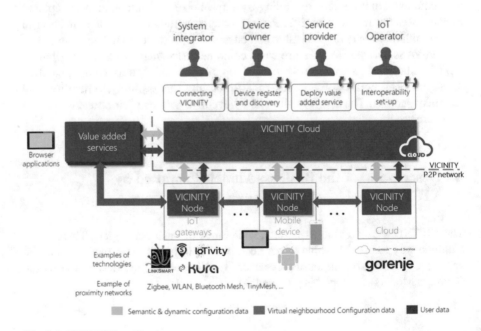

Fig. 3.4 VICINITY architecture

4.2 PaaS + Value-Added Service

VASs presented in this chapter have been developed by domain experts mainly in these four domains:

- Health
- Smart Parking
- Smart Energy
- Smart Buildings

The VASs implemented to demonstrate how realisation of cross-domain use case applications can be achieved with the help of the VICINITY IoT platform.

Data are collected for further processing, from devices of IoT infrastructures with which the VASs can be connected through the VICINITY platform. Furthermore, the realisation of such cross-domain VASs explores the potential for the evolution of new business models and value creation. These are enabled by exploiting the volume of semantically enhanced information stemming from the emerging IoT ecosystems connected to the IoT interoperability platform.

These services complement the offer of the PaaS. The VASs provide information about the sensors' data of an IoT infrastructure, after processing them, elaborating complex information the value of the data is enhanced. Data also could be extracted from other IoT infrastructures. They further reveal a business model potential/commercial exploitation of such a service (e.g., for application developers, service operators). The VASs developed in the context of the VICINITY platform can be further extended, adapted, and integrated to other businesses.

4.3 PaaS + Consultancy

Specialised consultancy services are provided to organisations willing to embrace the Digital transformation promised by the IoT. Some of the partners of VICINITY provide consultancy services to offer an integrated view of these sensors which could enable novel services to support organisations to make more strategic and informed decisions.

4.4 Data Market Place

VICINITY provides sovereignty to the owner of the data to decide what data to share and with whom to share. Rewarded mechanisms must be implemented to reward data owners either economically or with insights.

The Data Market Place opens new ways of commercialisation, under the VICINITY principle of user data sovereignty.

5   Business Model Design of Gorenje Group, A Digital Transformation Use Case

A digital business strategy is a strategy on how to adopt digital technologies and create business benefits in the ecosystem of people, assets, processes, and community-related systems. Digital business refers to business models, solutions, initiatives, and activities enabled and supported by digital technologies.

Gorenje Group is moving from being a traditional manufacturer towards the services-oriented organisation. This transition is impacting the whole company so a new set of skills, systems, processes, contents, and technologies is needed.

The purpose of the Gorenje Group's Digital Business Strategy is to create value by offering innovative digital business solutions for connected communities of their stakeholders and improving business efficiency using digital business solutions [40].

The related communities of stakeholders in the digital transformation process of the Gorenje Group are consumers, commercial and other business partners, financial partners, including shareholders, employees, and the wider social environment.

The Gorenje Group's digital business model as shown in Fig. 3.5 includes the following sets of digital business solutions:

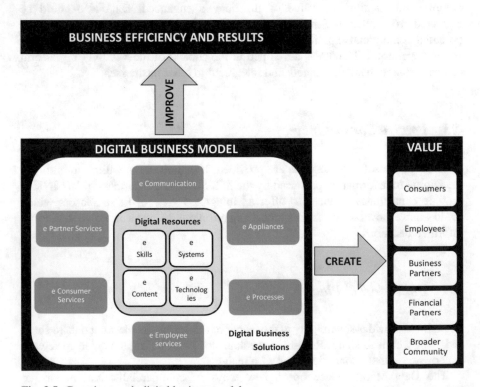

Fig. 3.5 Gorenje group's digital business model

- digital services for different connected communities,
- connected smart home devices,
- business processes backed by digital technologies.

To support the implementation of the digital business strategy, the Gorenje Group intends to continuously improve the following digital resources:

- digital skills,
- digital systems,
- digital content, and
- digital technologies.

The final objective of the Gorenje Group's digital business model is to create new values that are really understood and perceived by their most important groups of stakeholders: consumers, employees, business partners, financial partners, and the broader community. Once these new values are created, they can significantly contribute to the improved business efficiency, sales, financial results, and, consequently, to the successful digital business strategy execution.

Based on the digital business model, Gorenje Group is taking a holistic digital business transformation approach. On the organisational level, the company has established Digital Business Committee chaired by Chief Digital Officer who reports directly to the Managing Board. Within the digital business strategy, Gorenje Group has identified several projects that are grouped into logical programs in order to execute this strategy effectively. An example of Gorenje Group's Connected Appliances Project and its scope is presented in Fig. 3.6. The results of this project are illustrated in Fig. 3.7.

Projects are carried out by skilled project managers and teams of experts from different disciplines, such as R&D electronics, R&D appliances, IT, communication, IoT, digital marketing, etc. Gorenje Group has identified as necessary to combine its core business expertise with expertise from reliable external partners.

In order to achieve the goals from Gorenje Group's Digital Business Strategy, the company is using a comprehensive project portfolio management [41]. Recommended tactics for similar institutions that enter on a digital business transformation

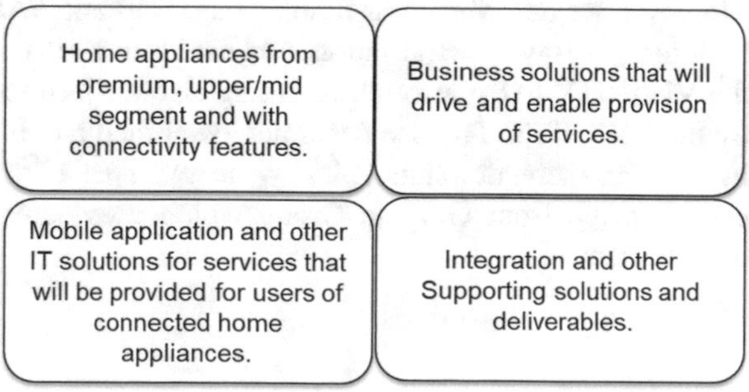

Fig. 3.6 Gorenje group's connected appliances project—scope

Home appliances from premium, upper/mid segment and with connectivity features.

Business solutions that will drive and enable provision of services.

Mobile application and other IT solutions for services that will be provided for users of connected home appliances.

Integration and other Supporting solutions and deliverables.

Fig. 3.7 Solutions developed within Gorenje group's connected appliances project

mission is a step by step approach which minimises risks and enables getting experience and building knowledge.

6 Conclusions

In this chapter, we present an overview of business models together with the impact that IoT has on them and the challenges identified. We conclude that the new era of IoT systems has a significant influence on business model strategies of the industry domain, as services and digital business models are arousing. Companies are adapting to this pave by transforming from traditional manufacturers towards services-oriented systems. Barriers and challenges that exist in this direction and IoT business models that are commonly adopted in order to overcome these barriers are presented in this chapter. As the IoT market is evolving, the industry is paving the way into a digital transformation as the interest in data analytic and insights from IoT devices and services is continuously increasing. In this context, new parameters are integrated in the whole process of a company's business plan. Moreover, we describe cross-domain use cases and VASs for the domains of smart buildings, energy, transportation, and health that are developed and integrated into the VICINITY IoT platform presenting also the business models that are proposed by the VICINITY. The use cases deployed facilitate in some identified challenges in each of these domains. Finally, an example of a digital transformation use case is given from Gorenje Group which uses comprehensive project portfolio management.

References

1. *The new thing: A silicon valley story* (2000). New York: W.W. Norton.
2. Gassmann, O. (2013). 55 pattern cards St. Gallen business model navigator. [St. Gallen]: BMI lab AG.
3. Shafer, S. M., Smith, H. J., & Linder, J. C.: The power of business models. *Business Horizons, 48*(3), 199–207.
4. What are some examples of a business model? https://www.aha.io/roadmapping/guide/product-strategy/what-are-some-examples-of-a-business-model
5. Turber, S., vom Brocke, J., Gassmann, O., & Fleisch, E. (2014). Designing business models in the era of Internet of Things. In M. C. Tremblay, D. VanderMeer, M. Rothenberger, A. Gupta, & V. Yoon (Eds.) *Advancing the impact of design science: Moving from theory to practice* (pp. 17–31). Cham: Springer.
6. How the Internet of Things changes business models. Harvard Business Review (2014). https://hbr.org/2014/07/how-the-internet-of-things-changes-business-models/
7. Pei Breivold, H., & Rizvanovic, L. (2018). Business modeling and design in the Internet-of-Things context. In *2018 IEEE 11th International Conference on Cloud Computing (CLOUD)* (pp. 524–531).
8. Smith, D. J. (2013). Power-by-the-hour: The role of technology in reshaping business strategy at Rolls-Royce. *Technology Analysis & Strategic Management, 25*(8), 987–1007.
9. Wolfe, J. (2019). Roomba vacuum maker iRobot betting big on the 'smart' home. https://www.reuters.com/article/us-irobot-strategy-idUSKBN1A91A5?il=0
10. Hodapp, D., Remane, G., Hanelt, A., & Kolbe, L. (2019). Business models for Internet of Things platforms: Empirical development of a taxonomy and archetypes. *In 2019 14th International Conference on Wirtshaftsinformatik.*
11. Dawson, C. (2019). Mitsubishi bets people will reveal their driving habits to insurers for a freebie. https://www.wsj.com/articles/mitsubishi-bets-people-will-reveal-their-driving-habits-to-insurersfor-a-freebie-1530853415
12. MacGillivray, C., & Reinsel, D. (2019). Worldwide global datasphere IoT device and data forecast, 2019–2023. Technical report, International Data Corporation (IDC).
13. Dimitrov, D. V. (2016). Medical Internet of Things and big data in healthcare. *Healthcare Informatics Research, 22*(3), 156–163.
14. Ghanbari, A., Laya, A., Alonso-Zarate, J., & Markendahl, J. (2017). Business development in the Internet of Things: A matter of vertical cooperation. *IEEE Communications Magazine, 55*(2), 135–141.
15. Chaudhari, S. S., & Bhole, V. Y. (2018). Solid waste collection as a service using IoT-solution for smart cities. In *2018 International Conference on Smart City and Emerging Technology (ICSCET)* (pp. 1–5). Piscataway: IEEE.
16. Doffman, Z. (2019). Cyberattacks on IoT devices surge 300% in 2019, 'measured in billions', report claims. https://www.forbes.com/sites/zakdoffman/2019/09/14/dangerous-cyberattacks-on-iot-devices-up-300-in-2019-now-rampant-report-claims/#f34504958926
17. Higginbotham, S. (2018). The internet of trash: IoT has a looming e-waste problem. *IEEE Spectrum: Technology, Engineering, and Science News, 17*. https://spectrum.ieee.org/telecom/internet/the-internet-of-trash-iot-has-a-looming-ewaste-problem
18. AboBakr, A., & Azer, M. A. (2017). IoT ethics challenges and legal issues. In *2017 12th International Conference on Computer Engineering and Systems (ICCES)* (pp. 233–237). Piscataway: IEEE.
19. Likar, M., & Miklav, A. (2016). Digital business strategy - planning and implementation. In *25th Annual Conference of the Slovenian Association for Quality and Excellence*, Portorož, Slovenia.
20. European Commission Buildings. (2017). https://ec.europa.eu/energy/en/topics/energy-efficiency/buildings

21. Economidou, M., Atanasiu, B., Despret, C., Maio, J., Nolte, I., Rapf, O., et al. (2011). Europe's buildings under the microscope. A country-by-country review of the energy performance of buildings. http://bpie.eu/wp-content/uploads/2015/10/HR_EU_B_under_microscope_study.pdf
22. Shan, M., Hwang, B.-G., & Zhu, L. (2017). A global review of sustainable construction project financing: Policies, practices, and research efforts. *Sustainability, 9*(12), 2347.
23. Ramprasad, B., McArthur, J., Fokaefs, M., Barna, C., Damm, M., & Litoiu, M. (2018). Leveraging existing sensor networks as IoT devices for smart buildings. In *2018 IEEE 4th World Forum on Internet of Things (WF-IoT)* (pp. 452–457).
24. Minoli, D., Sohraby, K., & Occhiogrosso, B. (2017). IoT considerations, requirements, and architectures for smart buildings-energy optimization and next-generation building management systems. *IEEE Internet of Things Journal, 4*(1), 269–283.
25. Spaced out: Perspectives on parking policy (2012). https://www.racfoundation.org/research/mobility/spaced-out-perspectives-on-parking
26. Mohanty, S. P., Choppali, U., & Kougianos, E. (2016). Everything you wanted to know about smart cities: The Internet of Things is the backbone. *IEEE Consumer Electronics Magazine, 5*(3), 60–70.
27. Silva, B. N., Khan, M., & Han, K. (2018). Towards sustainable smart cities: A review of trends, architectures, components, and open challenges in smart cities. *Sustainable Cities and Society, 38*, 697–713.
28. Lin, T., Rivano, H., & Le Mouël, F. (2017). A survey of smart parking solutions. *IEEE Transactions on Intelligent Transportation Systems, 18*(12), 3229–3253.
29. Papaioannou, T. G., Kotsopoulos, D., Bardaki, C., Lounis, S., Dimitriou, N., Boultadakis, G., et al. (2017). IoT-enabled gamification for energy conservation in public buildings. In *2017 Global Internet of Things Summit (GIoTS)* (pp. 1–6).
30. Pan, J., Jain, R., Paul, S., Vu, T., Saifullah, A., & Sha, M. (2015). An Internet of Things framework for smart energy in buildings: Designs, prototype, and experiments. *IEEE Internet of Things Journal, 2*(6), 527–537.
31. Avancini, D. B., Martins, S. G. B., Rabelo, R. A. L., Solic, P., & J. J. P. C. Rodrigues. A flexible IoT energy monitoring solution. In *2018 3rd International Conference on Smart and Sustainable Technologies (SpliTech)* (pp. 1–6).
32. Militello, A., Borra, M., Bisegna, F., Burattini, C., & Grandi, C. (2016). Smart technologies: Useful tools to assess the exposure to solar ultraviolet radiation for general population and outdoor workers. In *18th Italian National Conference on Photonic Technologies (Fotonica 2016)* (pp. 1–4).
33. Farahani, B., Firouzi, F., Chang, V., Badaroglu, M., Constant, N., & Mankodiya, K. (2018). Towards fog-driven IoT eHealth: Promises and challenges of IoT in medicine and healthcare. *Future Generation Computer Systems, 78*, 659–676.
34. Eurostat: A look at the lives of the elderly in the EU today. https://ec.europa.eu/eurostat/cache/infographs/elderly/index.html
35. Eurostat statistics explained. Overweight and obesity, BMI statistics. https://ec.europa.eu/eurostat/statistics-explained/index.php/Overweight_and_obesity_-_BMI_statistics
36. Commission of the European Communities. (2007). White paper on a strategy for Europe on nutrition, overweight and obesity related health issues.
37. Official Journal of the European Union. (2014). Council conclusions on nutrition and physical activity, c 213/1-6.
38. Greece profile of health and well-being (2016). World Health Organization.
39. Hellenic Statistical Authority. (2011). Population-housing census web site. http://www.statistics.gr/en/2011-census-pop-hous
40. Debeljak, Ž. (2015). Gorenje group: Digital business strategic guidelines. Technical report, Gorenje Group's internal document.
41. Inc Project Management Institute. (2017). *The standard for portfolio management* (4th ed.). Newtown Square: Project Management Institute.

Chapter 4
Methods and Tools for Validation and Testing

Johannes Kölsch, Yajuan Guan, and Christoph Grimm

1 IoT Development Process

IoT development process requires various technologies and disciplinary knowledge that involves hardware, software, middleware, platform, standardization, etc. Before starting the implementation of an IoT project, research the appropriate IoT development tools, platforms, and applications that could help to successfully develop the IoT product. In order to make sure that the developed IoT project operates correctly from a technical perspective, intensive and a variety of testing should be done based on simulation platforms and testbeds to evaluate the performance, identify bugs, and improve the core components and integration before the actual deployment on-site. The simulation platforms and testbeds are able to emulate different use cases and testing scenarios, as well as providing the capability to test the IoT core components and prototypes with extreme situations and edge-cases without resulting in the real system crash.

This chapter aims to provide a detailed introduction and comparative analysis for the commonly used IoT validation and testing methods and tools. The rest of the chapter is organized as follows: Detailed introductions regarding IoT simulation tools and IoT testbed are provided in Sect. 2 and 3, respectively. Section 4 introduces an IoT simulation example of the VICINITY project. Section 5 concludes the chapter.

J. Kölsch (✉) · C. Grimm
TU Kaiserslautern, Kaiserslautern, Germany
e-mail: koelsch@cs.uni-kl.de; grimm@cs.uni-kl.de

Y. Guan
Department of Energy Technology, Aalborg University, Aalborg Øst, Denmark
e-mail: ygu@et.aau.dk

© Springer Nature Switzerland AG 2021
C. Zivkovic et al. (eds.), *IoT Platforms, Use Cases, Privacy, and Business Models*,
https://doi.org/10.1007/978-3-030-45316-9_4

2   IoT Simulation

The IoT and smart grid domains are made of hundreds, or even thousands of devices that are deployed in highly dynamic and unreliable environments [1]. Such applications are generally supported through clouds where data is stored and handled by processing systems. Such systems need to be adapted at runtime to take the environment changes into account [2]. A major challenge is how to develop such adaptive IoT software systems, it has to be reliable, ensuring that the devices will perform their tasks under different conditions, that all devices are ready to carry out their tasks when needed and of course that the different devices and systems are interoperable [3, 4]. This is impractical for empirical evaluation on a real network. Therefore it is common to use simulations in the domain of IoT to verify the properties and behaviors of the system.

There are many different kinds of simulations. For event-driven simulations, the global state of the system changes only as a function of events meaning that the value of the global time sometimes jumps by a large amount, at other times it is not changing while going through several events. They are rather unpractical for IoT and smart grid domains.

A discrete event simulator focuses on the processes in a system. The global state of the system changes as a function of time, and is affected by events which can be generated by events within the system or by events generated externally.

Another type is trace-driven simulations. They have the advantage that the implementation of trace-driven models is very similar to the system being modeled and that performance characteristics can be measured at the same time and compared with the output of the simulator. Traces also retain the correlation and interference effects in the workload which makes traces very credible compared to random sequences using an assumed distribution. The drawbacks with trace-driven simulations especially in regard to IoT are that they require a more detailed simulation of the system and together with the fact that traces get obsolete faster than other forms make it very time-consuming. Additionally to validate a result of several traces that take up a lot of space. Therefore trace-driven simulations seem unappealing for huge and complex systems in the IoT.

Most real-world scenarios are very complex and are linked to many other different domains where it would be advantageous to use different simulation methods for different parts of the system. Using different methods of modeling and simulation to overcome the drawbacks of individual approaches and granting higher flexibility is the aim of the Multi-method Simulation Modeling.

In the next few subsections, the most known and used simulation tools in the domain of IoT will be briefly presented.

2.1   NS-2

Network Simulator 2 (NS-2) is a discrete event simulation tool and has proved its worth in research of dynamic communication networks. NS-2 was developed in the year 1989 [5]. It consists of two key languages: C++ and Object-oriented Tool Command Language (OTcl). While C++ is mainly used for implementing various protocols and extending simulation libraries, OTcl scripts are used for setting up the simulation configurations, network topology, and the scenarios for the simulation [5].

NS-2 is mostly used because it provides support for various protocols and allows for high flexibility in building the network topology. Nevertheless, it has some heavy drawbacks, among them the non-existence of sensing models which makes the parameters used in the simulation completely different to real-world sensor network scenarios [5]. Furthermore, NS-2's performance in Wireless sensor network (WSN) simulations is bad in regard to debugging, tracing, speed and also needs a lot of memory to simulate which makes it unpractical for WSN and IoT scenarios. In NS-2 it is impossible to create a graphical editor for simulations. This makes it hard for newcomers to use the system.

2.2   NS-3

The Network Simulator 3 (NS-3) is also a discrete event simulator launched in June 2008. NS-3 is a new simulator not supporting any API's belonging to NS-2. The main difference in NS-3 is that all programs are written in C++ which enables simulating models that are closer to real-world sensor networks. Still in NS-3 graphical editors are not possible which many users have expressed is a problem, making it too time-consuming to get into it [5].

2.3   OPNET

Optimized Network Engineering Tools (OPNET), a discrete event simulation tool that was created by OPNET Technologies Inc. It is considered to be one of the most powerful and user-friendly collections of toolsets to create and test large network environments. Reasons for this are for one its large protocol model library (including IPv6, MIPv6, WiMAX, QoS, Ethernet, MPLS, OSPFv3, and many others). Secondly the graphical editor when defining network topology, allowing for quick and natural use of drag and drop actions to create high-end network environments. OPNET also comes with a powerful command-line simulation debugger and also provides a graphical debugger and automatic animations which have been proven to simplify model development and make it more accessible [6, 7].

The most notorious drawbacks when using OPNET are that OPNET models are always of fixed topology, which means that OPNET is inflexible in simulations in WSN (which applies in IoT as well) because of the different environments and their use of different protocols and topologies [6]. Furthermore, OPNET models are difficult to generate by program because the models are stored in a proprietary binary file format.

2.4 IOTSIM

IoTSim is built upon CloudSim, which is a popular extensible simulation toolkit that concentrates on modeling and simulation of cloud computing environments and supports modeling of virtual machines on simulated nodes. IoTSim adds IoT application model support and enables the processing of IoT data using the MapReduce framework. The simulator also allows modeling and simulating network usage between storage and processing virtual machines [8]. The drawbacks of using IoTSim are caused by the limitations of MapReduce in regard to IoT applications that have real-time and low-latency requirements [9].

2.5 J-Sim

JavaSim is a component-based simulation environment, implemented in Java. The fact that it uses Java makes it strong in debugging and model development; however, it also weakens the simulation performance and makes it not possible to reuse existing real-life protocols written in C as simulation models.

It is not possible to implement graphical editors for J-Sim which does provide one but the format used (XML) makes it hard to use. J-Sim seems to have too many drawbacks as the development of J-Sim has stopped since 2004 [7].

2.6 OMNeT++

OMNeT++ (Objective Modular Network Testbed in C++) is a simulation framework that offers many tools and extensions for network/IoT simulations. OMNeT++ provides a modular simulation environment with strong GUI support and an embeddable simulation kernel. It is public-source and benefits from a large community which also provides numerous plug-ins of different natures (from overlays to complete model frameworks). Additionally, it contains an ecosystem of simulation module libraries focusing, to name a few, on Internet protocols, wireless networks, overlay networks, and vehicular networks, etc. [10].

OMNeT++ is designed to simulate discrete event systems, but the primary application area is the simulation of communication networks [10]. The current version of OMNeT++ 5.4.1. is compatible to run on different operating systems like Windows, Linux, and MAC OS X. It also includes an Eclipse base IDE environment which enables programming and debugging of modules. Additionally, it allows for graphical and textual editing of NED (Network Description) files. NED lets the user declare simple modules and connect and assemble them into compound modules, in which the OMNeT++ elements are composed of [5]. One module can represent a network protocol or hardware components like a sensor and communicates with other modules by exchanging messages.

It is possible to run simulations directly from the integrated environment, as a C++ application or as a standalone application. Furthermore, it is possible to set up reusable datasets with different parameter settings to run several simulations with. The results of the simulation are written into text-based files but can be visualized by processing them with other tools, for example, Matlab. Additionally, the OMNeT+ simulation kernel is capable of logging various events during simulation: scheduling and canceling self-messages, sending messages, displaying changes, module and connection creation and deletion, user log messages, etc. [5].

Together with all the information collected in the event log file and results file, OMNeT++ offers different tools that make use of its graphical runtime environment and allow for automatic animation, module output windows, object inspectors, and different visualizations of, for example, the messages exchanged between modules or the results, etc. [5].

2.7   Matlab

Matlab offers many analytic and visual tools for various domains. (for example, it has toolboxes for Aerospace, Fuzzy Logic, Symbolic Computations, Statistics, Communications,...) Many frameworks are being used in conjunction with Matlab, the most known being Simulink and ThingSpeak. The main advantage of using Matlab is that it is easy to use and enables simple analysis of workflows of different domains, for example, Robert S. Mawrey has used ThingSpeak and Matlab to simulate an IoT system that makes a tide forecast dependent on current tide and wind level. He concluded that this could easily be done for other domains, namely for power load forecasting which is important for smart grids [11].

2.8   SimIoT

In [12] S. Sotiriadis, N. Bessis, E. Asimakopoulou, and N. Mustafee describe a approach called SimIOT, which uses the cloud to execute the backend operations to improve the performance of the simulator. This simulator is based on SimIC [13], which is a discrete event simulation toolkit using an inter-cloud facility. It is based on three-level: User level, SimIOT level, and SimIC level (the last two are in the cloud). All data, which is generated by sensors or other devices on the User level, is sent to the cloud with automated requests. In the SimIOT level the communication broker translates this information and forwards an information processing request to the SimIC cloud entities. The performance and tracing operations are calculated in the SimIC level. Performance includes, for example, execution time and service latencies, tracing includes logging of events.

The big advantage of this approach is the simulation calculation in the cloud. It can scale up and leads to high simulation performance.

2.9 MAMMOTH

V. Looga, Z. Ou, Y. Denga, and A. Ylä-Jääski present in [14] another concept of modeling the Internet of Things. Emulation is used there instead of simulation, because in emulation the same software is used on the nodes as on real hardware, so you can test the software directly on the hardware. Against this, in simulation, a simplified software is used to note only some specific aspects. Also the correct timing of events in the simulation could leave some software problems of message timing uncovered. MAMMOTH is such a large-scale emulator, which should emulate thousands of nodes on a single (virtual) machine. It presumes three scenarios: Mobile devices connected to a base station in a star topology, a Wireless Sensor Network connected to a base station, and a number of constrained devices connected to proxies, which in turn are connected to a backend. To make the network traffic most realistic, connections between nodes can use the profile of General Packet Radio Service (GRPS) or Wireless Personal Area Networks (WPAN). To simulate UDP or TCP traffic, a Netfilter-type traffic scheduler is used. A proprietary Java-based node emulator is used to run MAMMOTH.

All in all this technique seems interesting, but for the house simulation, it is not the intended solution, because the developed software should not run in reality.

2.10 IoT Node

Another solution for the simulation of interconnected IoT devices is [15] by G. Brambilla, M. Picone, S. Cirani, M. Amoretti, and F. Zanichelli. IoT nodes represent generic smart objects with a mobility model, one or more network models, and an energy model. The energy model describes the physics of the node. The network models characterize the networks of the node including delays and package losses. Finally with the energy model, the energy consumption behaviour of a node can be specified.

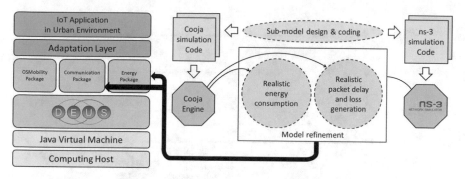

Fig. 4.1 Layers of the IoT node [15]

The architecture of IoT nodes can be seen in Fig. 4.1. The top layer represents the applications, which should be tested and the Adaption Layer is responsible for the communication of the nodes. The mobility model uses OSMobility, an extension of DEUS [16], which is a discrete event simulation environment, to simulate the motion. Cooja [17] and ns-3 [18], which are two other simulation tools, are used for the network model and Cooja is also used for the energy consumption model.

To test this approach a use case of an urban smart parking scenario is presented. The scenario consists of sensor nodes, which detect cars in parking lots, gateways, which collect the data from the sensor network and vehicles, which move around and contact the gateway to get a parking lot. The experiments show that the simulation works with thousands of nodes at the same time.

2.11 Multi-Level Simulation

To improve scalability and real-time execution for big IoT environments G. D'Angelo, S. Ferretti, and V. Ghini introduce another simulation approach [19]. It uses Parallel and Distributed Simulation (PADS) together with multi-level simulation.

PADS is a simulation running on more than a processor to enhance execution speed, so that every processing unit manages only a part of the simulation. To execute this, the simulation model is separated into different Logic Processors (LP), so that the communication between different LPs is as low as possible. These LPs are connected directly to the processing units. This separation is only valid, if the result is exactly the same as in sequential simulation. This is done with synchronization algorithms.

To make it possible to a simulation with a large number of nodes, a multi-level architecture is used to extend the PADS. In this case, a "high level" simulator (i.e.,

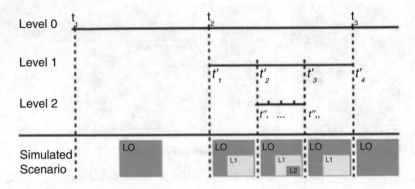

Fig. 4.2 Multi-level simulation [19]

GAIA/ARTIS[1]) is used. For "middle" and "low level," other simulators for fine-grained level of detail are used (e.g., OMNeT++ [20] or ns-3 [18]). The Change between the levels can either be automatic or triggered. In Fig. 4.2 can be seen that at level 0 the main scenario is executed. If at a certain timestamp (t_2 in the figure) a part of the scenario needs more details, it will be modeled at level 1. The other part of the scenario is still executed at level 0. The more detailed execution at level two can also be extended to one more level for more details of a part of the scenario. If at another timestamp (t_4' in the figure) no more details are needed in level 1, all components of level 1 are transformed back into the level 0 simulation.

With the multi-level simulation, the total number of nodes used does not change, but this approach enables better scalability because of the different levels of detail. If fewer details are needed, the simulation can be executed at lower levels and less unnecessary details are simulated. On the other hand with the use of multi-level, there are many more possibilities for mistakes, so verification and validation techniques are very important.

G. D'Angelo presents a use case for providing smart services to territories like cities or decentralized areas focused on "smart shires." This means, that it is possible, to manage resources like nature, humans, buildings, and infrastructure in a sustainable way to the environment, so that with smart and cheap services the life of citizens and tourists is improved. With the connection of cheap sensors their data can be collected. An example for this is the "km 0" phenomenon. The idea is to use local, seasonal, and not genetically modified food, which is not easy to buy. With the service, costumers can be notified, if a subscribed product is available by a local producer, get all important information about it and can book specific items. For this use case multi-level simulation can be used. Level 0 simulates the whole smart territory, where products are created, interests are subscribed and are moved. This can be implemented with PADS. Wireless communication and refined interactions

---

[1]http://pads.cs.unibo.it/doku.php?id=pads:download.

are simulated at level 1. Each simulation step of level 0 is decomposed into multiple substeps at level 1.

2.12   Parallel Discrete Event System (PDEVS)

In [21] J. Kölsch, A. Ratzke, and C. Grimm describe a simulation approach for IoT use cases using discrete computing parts and continuous-time dynamic parts. The core simulation uses Parallel Discrete Event Systems (PDEVS). Parallel DEVS is based on Finite State Automata and uses some concepts of Discrete Event Simulation [22]. They can process many internal and external events at the same time. Also a third state change function is added with the time advance function which changes the state after a given time. The core Simulation also routes inputs and outputs of the models and schedules autonomous events. The whole architecture of the framework can be seen in Fig. 4.3.

The used models can be divided into two classes:

1. **Atomic models**: They are the smallest models representing an object of the simulation and can be described with their state change functions for internal, external, and autonomous events.
2. **Network models**: They are composed of multiple atomic models and sometimes other network models and are responsible for the routing of the atomic models.

To use a multi-level simulation, some atomic models should be exchanged with network models with more details. The atomic models approximate the behavior at a lower level which improves the performance. The Network model is dynamically interchangeable at a higher level of detail, but has caused by the complexity of smaller timesteps. These two models are designed as an HierarchicalAtomic, so they can switch from a detailed network model to a faster (high level) atomic model

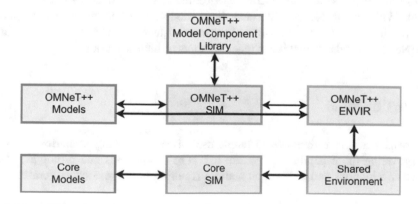

Fig. 4.3 Multi-level simulation [21]

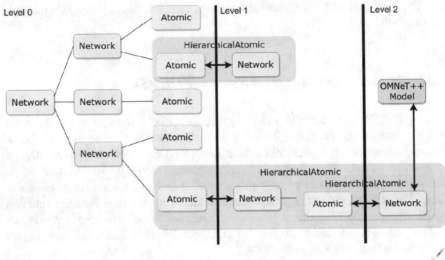

Fig. 4.4 The model tree of hierarchical levels [21]

during runtime. These models can be composed in a tree-like structure, which can be seen in Fig. 4.4. It is recommended, to use a bottom-up procedure, where the level with the most details is designed first in its whole and after that, the partitioning into smaller leaves of the tree can be done.

The network simulator OMNeT++ [20] is used for the simulation. With the INET [2] extension this tool is used for the communication, energy consumption, and movement of Cyber-Physical Systems. To connect the OMNeT++ simulation with the core simulation described above, a Shared Environment is used, which provides the functionalities of both simulators. The sequential scheduler of OMNeT++ is extended to attend events for both simulators, so that synchronization events are produced to determine which one of the simulators executed before the other one. These events are sent to the Shared Environment, which advances the simulators state. With an OMNeT++ interface, both models can communicate with each other. It implements characteristics of the event listener of the core framework and OMNeT++ models, so that it can react to inputs of both scenarios.

3 IoT TestBed

Testbeds are able to emulate different use cases and testing scenarios, as well as providing the capability to test the IoT core components and prototypes with extreme situations and edge-cases without resulting in the real system crash. Two

---

[2]https://inet.omnetpp.org/.

open IoT testbeds which can be accessed by researchers for experiments are introduced in this subsection [23].

3.1 FIT IoT-LAB

FIT IoT-LAB[3] offers a large-scale testing infrastructure which can test wireless sensors and various communicating devices.

FIT IoT-LAB offers complete control of connected nodes and direct access to the gateways, thereby remotely monitoring the energy consumption of the connected notes and network-related metrics. The facility allows easy experiments deployment, together with information collection, data analysis, and primary evaluation. The flexibility of FIT IoT-LAB allows the developers to define different types of testbeds with various nodes, architectures, and environments, thereby covering a series of domains and use cases.

FIT IoT-LAB testbeds are located in different sites in France which provides forward access to 1786 wireless sensor nodes as listed below. Figure 4.5 shows the deployment at Inria Lille[4]:

- Inria Grenoble - Rhône-Alpes (640),
- Inria Lille - Nord Europe (293),
- ICube - Strasbourg (400),
- Inria Saclay - Île-de-France (264),
- Institut Mines - Télécom - Paris (160),
- CITI Lab - Lyon (29).

Fig. 4.5 IoT-lab deployment at Inria Lille

https://www.iot-lab.info/deployment/lille/

---

[3]FIT IoT-LAB: https://www.iot-lab.info/what-is-iot-lab/.

[4]FIT IoT-LAB at Inria Lille https://www.iot-lab.info/deployment/lille/.

Fig. 4.6 WSN430 v1.4

https://www.iot-lab.info/hardware/wsn430/

The FIT IoT-LAB hardware infrastructure consists of a range of FIT IoT-LAB nodes. The power and connectivity of FIT IoT-LAB nodes are supplied by a global networking backbone.

The WSN430 open node, as shown in Fig. 4.6, is one of the IoT-LAB boards[5] which is developed particularly for FIT IoT-LAB testbed. It is based on a low power MSP430-based platform, with a fully functional ISM radio interface and a set of standard sensors.

3.2 Japan-Wide Orchestrated Smart/Sensor Environment

JOSE[6] (Japan-wide Orchestrated Smart/Sensor Environment) testbed offers a flexible and customizable experimental environment which offers services and a data-sharing mechanism. The developers can use JOSE network and cloud facilities across multiple data centers in Japan. JOSE can be used to test new IoT technologies using the network and cloud facilities, and to establish secure field experimental environments. There are 1250 computers (up to 20,000 virtual machines) running at five data centers in Japan. Each computer has a network connection at up to 10 Gbps.

JOSE offers an all-purpose Linus operating system which has full network connectivity, and security and software frameworks.

The testbed comprises massive computers and large-capacity data storages that connected by broadband networks (max 100Gbps), therefore it is allowed to develop an information and communications technology (ICT) environment. The network and computer connectivity and configuration are able to be easily modified by using JOSE Software Defined Infrastructure (SDI) function. The environment offers

---

[5]IoT-LAB boards: https://www.iot-lab.info/hardware/wsn430/.

[6]JOSE: https://testbed.nict.go.jp/jose/english/index.html.

security, independence, and isolation features for the users. The developers can customize their assigned environments and develop their own systems.

4 IoT Simulation Examples of the VICINITY Project

In order to further ensure that the VICINITY platform operates correctly from a technical perspective prior to deployment at the pilot sites. Intensive and iterative Lab tests have been conducted on the Hardware-in-the-Loop/experimental platform in four different laboratories (AAU Microgrid-IoT lab, CERTH/ITI Smart House, ATOS IoE lab, and UNIKL lab). Two kinds of testing plans (edge-cases and internal point cases) were designed and performed based on Edge Case Testing Methodology to have a good coverage over the range of values.

The internal testing points track the use cases defined for the pilot sites to verify the functional performance of VICINITY core components, adapters, and VAS, and ensure the expected operation. The tests cover mobility, building, energy, and eHealth domains and refer to privacy, GDPR VAS, LoRa, and FIWARE-compliant devices, an Omnet++ network simulator and the homomorphic encryption method [24].

The following subsections introduce two internal testing scenarios as examples for the IoT simulation and Laboratory testing.

4.1 Hardware-in-the-Loop (HiL) Simulation Platform-Based Testing at Aalborg University Microgrid-IoT Laboratory

As observed in Fig. 4.7, Matlab/Simulink, dSPACE 1006 HiL simulation platform, and three DC/AC power inverters are used to emulate a Photovoltaic (PV)/wind turbine/battery hybrid residential microgrid which assumes to deliver electricity to the GORENJE smart appliances and three parking slots in case of electrical vehicles (EV).

The GORENJE smart refrigerator, GORENJE smart oven, and PlacePod smart parking sensors are registered to the VICINITY virtual neighborhood manager (VNM) in Device catalogue through their adapters, VICINITY agents, and VICINITY gateway API.

A LabVIEW-based residential microgrid energy management system and a user interface are developed to achieve optimized control for energy resources and local loads Fig. 4.8, and to calculate the real-time EV charging price by considering the forecasts of renewable energy outputs and load consumption. The residential microgrid energy management system is registered in VICINITY VNM through a Python-based adapter as a VAS. The residential microgrid energy management system VAS subscribes to the events published by GORENJE smart appliances and

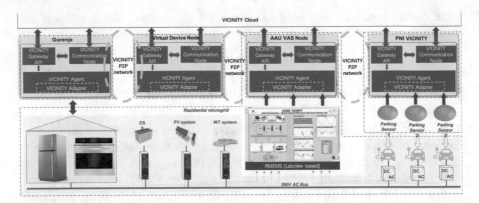

Fig. 4.7 IoT infrastructure based on hardware-in-the-loop simulation platform at AAU microgrid-IoT laboratory [24]

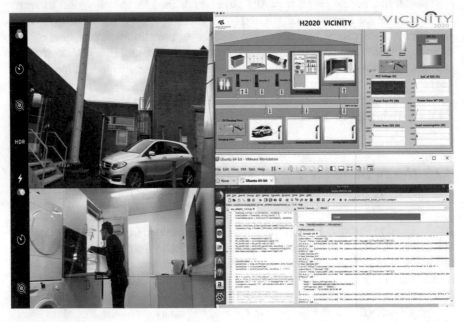

Fig. 4.8 Real-life testing scenario for smart parking and residential load scheduling in AAU Microgrid-IoT laboratory [24]

PlacePod smart parking sensors, and publishes events with the data of renewable energy outputs, load consumption, state of charge of batteries, real-time EV charging price, and puts actions, for instance, "start baking" or "fast freeze" to GORENJE smart appliances, according to the load scheduling commands.

4.2   TU Kaiserslautern Hardware-in-the-Loop Simulation Framework

At TU Kaiserslautern the approach from Sect. 2.12 was used to build up a HiL simulation lab. The use case simulated in the lab is an extension of the VICINITY pilot site in Tromsø (see Fig. 4.9. Most of the use cases, as described in Chap. 3 can be covered with the simulation.

To illustrate the applicability and performance of the developed multi-level simulator, we modeled and simulated a smart energy use case. This particular use case describes a smart energy scenario within a city. The city has a photovoltaic system and a windmill as power suppliers and a parking lot and a couple of houses as consumers. Electric vehicles can move inside the city and the parking lot. By using a smart parking service through a mobile app, users of the system can request to reserve their parking slot of choice within the participating parking facilities. The availability of the parking slots is then displayed through the mobile app as well as through the optical indicators located on the respective parking slots for random people that do not participate in the smart parking service.

The described scenario has been modeled and simulated using the proposed approach at three distinct levels of abstraction: The first two higher levels have been implemented only using classes provided by the core simulator. The third (lowest) level has been implemented with OMNeT++ 5.4.1[7] and its INET extension 4.0.[8]

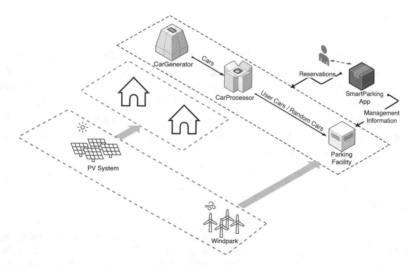

Fig. 4.9 Smart energy use case—highest level of abstraction [21]

---

[7]https://omnetpp.org/software/2018/06/29/omnet-5-4-1-released.html.

[8]https://inet.omnetpp.org/2018-06-28-INET-4.0.0-released.html.

The specific solution for this use case has used 3 levels of simulation. The first level has been used to generate abstract information on the general movement of simulated entities and communication between them. Furthermore the energy structures around the use case have been modeled. The second level has served as a space division for the lower level. It produced more detailed information about the movement that has been used as a basis to dynamically activate the different parts of the lowest level—Level 2. Level 2 has then used the powerful network simulator OMNeT++ with the INET framework to simulate the details of a smart parking service, the movement of users, the communication between them, the charging behaviour of cars and the environment.

With respect to interoperability, the simulator has proven to fulfill the requirements for an IoT simulator. This is achieved through the following capabilities:

1. Dynamic switching between models at different levels of abstraction
2. Spreading multiple simulation engines across the model tree
3. Modeling and simulation of mobile system entities and their communication through the OMNeT++ integration

The approach proposes interconnection between models at different abstraction levels within the *discrete event simulation framework* and has been demonstrated on a smart parking use case of the *VICINITY* pilot site in Tromsø together with energy considerations [21].

5 Conclusion

In this chapter, we have presented the IoT simulation and validation tools. First, we have introduced and compared the most known and commonly used IoT simulation tools. Choosing the appropriate simulation tools and methods depends on the need of each project. Secondly, we have presented two open IoT Testbeds which can be accessed by researchers to test the IoT core components and prototypes. Two testing scenarios of the VICINITY project are introduced as examples for the IoT simulation and Laboratory testing in the last section.

References

1. Costa, B., Pires, P. F., Delicato, F. C., Li, W., & Zomaya, A. Y. (2016). Design and analysis of IoT applications: A model-driven approach. In *2016 IEEE 14th International Conference on Dependable, Autonomic and Secure Computing, 14th International Conference on Pervasive Intelligence and Computing, 2nd International Conference on Big Data Intelligence and Computing and Cyber Science and Technology Congress (DASC/PiCom/DataCom/CyberSciTech)* (pp. 392–399).
2. Nogueira, J., Bhattacharya, S., & Luqi. (2000). A risk assessment model for software prototyping projects. In *2013 9th International Conference on Collaborative Computing: Networking, Applications and Worksharing (CollaborateCom), Los Alamitos* (p. 28). Washington: IEEE Computer Society. https://ieeexplore.ieee.org/author/37283803900

3. Al-Fuqaha, A., Guizani, M., Mohammadi, M., Aledhari, M., & Ayyash, M. (2015). Internet of Things: A survey on enabling technologies, protocols, and applications. *IEEE communications Surveys & Tutorials, 17*(4), 2347–2376.

4. Hussein, M., Li, S., & Radermacher, A. (2017). Model-driven development of adaptive IoT systems. In *MODELS (Satellite Events)* (pp. 17–23).

5. Nayyar, A., & Singh, R. (2015). A comprehensive review of simulation tools for wireless sensor networks (WSNS). *Journal of Wireless Networking and Communications, 5*(1), 19–47.

6. Xian, X., Shi, W., & Huang, H. (2008). Comparison of OMNET++ and other simulator for WSN simulation. In *2008 3rd IEEE Conference on Industrial Electronics and Applications* (pp. 1439–1443). Piscataway: IEEE.

7. Varga, A., & Hornig, R. (2008). An overview of the OMNET++ simulation environment. In *Proceedings of the 1st International Conference on Simulation Tools and Techniques for Communications, Networks and Systems & Workshops* (p. 60). ICST (Institute for Computer Sciences, Social Informatics and Telecommunications Engineering).

8. Zeng, X., Garg, S., Strazdins, P., Jayaraman, P. P., Georgakopoulos, D., & Ranjan, R. (2016). IoTSim: A cloud based simulator for analysing IoT applications. *Journal of Systems Architecture, 02*.

9. Zeng, X., Garg, S. K., Strazdins, P., Jayaraman, P. P., Georgakopoulos, D., Ranjan, R. (2017). IoTSim: A simulator for analysing IoT applications. *Journal of Systems Architecture, 72*, 93–107. Design Automation for Embedded Ubiquitous Computing Systems.

10. Förster, A., Minkenberg, C., Herrera, G. R., & Kirsche, M. (2015). Proceedings of the 2nd OMNET++ community summit, IBM research - Zurich, Switzerland, September 3–4, 2015. *CoRR*, abs/1509.03284.

11. Pavalkis, S. (2018). Overview of current sysML/UML and MATLAB/simulink ® integration use case and implementations. https://blog.nomagic.com/overview-of-current-sysml-uml-and-matlab-simulink-integration-use-cases-and-implementations/

12. Sotiriadis, S., Bessis, N., Asimakopoulou, E., & Mustafee, N. (2014). Towards simulating the Internet of Things. In *2014 28th International Conference on Advanced Information Networking and Applications Workshops* (pp. 444–448).

13. Sotiriadis, S., Bessis, N., Antonopoulos, N., & Anjum, A. (2013). SimIC: Designing a new inter-cloud simulation platform for integrating large-scale resource management. In *2013 IEEE 27th International Conference on Advanced Information Networking and Applications (AINA)* (pp. 90–97).

14. Looga, V., Ou, Z., Deng, Y., & Ylä-Jääski, A. (2012). Mammoth: A massive-scale emulation platform for Internet of Things. In *2012 IEEE 2nd International Conference on Cloud Computing and Intelligence Systems* (vol. 3, pp. 1235–1239).

15. Brambilla, G., Picone, M., Cirani, S., Amoretti, M., & Zanichelli, F. (2014). A simulation platform for large-scale Internet of Things scenarios in urban environments. In *Proceedings of the First International Conference on IoT in Urban Space, URB-IOT'14* (pp. 50–55). ICST (Institute for Computer Sciences, Social-Informatics and Telecommunications Engineering).

16. Amoretti, M., Agosti, M., & Zanichelli, F. (2009). Deus: A discrete event universal simulator. In *Proceedings of the 2nd International Conference on Simulation Tools and Techniques* (pp. 58:1–58:9). ICST (Institute for Computer Sciences, Social-Informatics and Telecommunications Engineering).

17. Eriksson, J., Finne, N., & Voigt, T. (2006). Cross-level sensor network simulation with COOJA. In *First IEEE International Workshop on Practical Issues in Building Sensor Network Applications (SenseApp 2006)*. Swedish Institute of Computer Science.

18. Riley, G., & Henderson, T. (2010). *The ns-3 network simulator* (pp. 15–34). Berlin: Springer.

19. D'Angelo, G., Ferretti, S., & Ghini, V. (2016). Simulation of the Internet of Things. In *2016 International Conference on High Performance Computing Simulation (HPCS)* pp. 1–8.

20. Wehrle, K., Günes, M., & Gross, J. (2010). *OMNeT++* (pp. 35–59). Berlin: Springer.

21. Kölsch, J., Ratzke, A., & Grimm, C. (2019). Co-simulating the Internet of Things in a smart grid use case scenario. In *2019 7th Workshop on Modeling and Simulation of Cyber-Physical Energy Systems (MSCPES)* (pp. 1–6).
22. Tendeloo, Y. V. & Vangheluwe, H. (2018). Discrete event system specification modeling and simulation. In *2018 Winter Simulation Conference (WSC)* (pp. 162–176).
23. Chernyshev, M., Baig, Z., Bello, O., & Zeadally, S. (2018). Internet of Things (IoT): Research, simulators, and testbeds. *IEEE Internet of Things Journal, 5*(3), 1637–1647.
24. Guan, Y., Čolić, N., Feng, W., Fernandez, D., Flsveen, F., Filosofov, D., Guerrero, J., Heinz, C., Horniak, M., HovstÃ, A., Koutli, M., Koelsch, J., Oravec, V., Palacios-Garcia, E., Poljakov, G., Tryferidis, T., Theologou, N., Vásquez, J., & Vanya, S. (2018). Report on VICINITY test-bed deployment, including validation, parameterization and testing.

Chapter 5
Ontologies for IoT Semantic Interoperability

Andrea Cimmino, Alba Fernández-Izquierdo, María Poveda-Villalón, and Raúl García-Castro

1 Introduction

Nowadays, due to the growth of the available IoT objects on the Web there is an increasing demand for transparent discovery and access of data published by such objects in IoT infrastructures [16].

The concept of interoperability refers to the ability of a system to transparently interact with others that are also interoperable [17]. There are several layers of interoperability, each of which aims at solving different problems: technical, syntactic, and semantic layer.

The technical interoperability addresses the way data is accessed, and thus, the access methods and the endpoints that the IoT infrastructures implement.

The syntactic interoperability addresses the way data is represented, and thus, the formats used to express the data.

Finally, *the semantic interoperability* addresses the form under which data is expressed, and thus, the models used to represent such data.

These three layers of interoperability build a hierarchical environment in which the transparent interaction among systems is enabled. Once an interoperable environment is built, i.e., systems that belong to the same environment implement the different layers of interoperability, transparent discovery and distributed access to the services of IoT infrastructures can be implemented on top of such environment [24]. The described interoperability framework is depicted in Fig. 5.1.

Semantic web technologies are the pillars to implement the aforementioned layers of interoperability [26]. On the one hand, the semantic web technologies are W3C standards, which make them reliable and trustworthy [18]. On the other hand,

A. Cimmino (✉) · A. Fernández-Izquierdo · M. Poveda-Villalón · R. García-Castro
Universidad Politécnica de Madrid, Madrid, Spain
e-mail: cimmino@fi.upm.es; albafernandez@fi.upm.es; mpoveda@fi.upm.es; rgarcia@fi.upm.es

© Springer Nature Switzerland AG 2021
C. Zivkovic et al. (eds.), *IoT Platforms, Use Cases, Privacy, and Business Models*,
https://doi.org/10.1007/978-3-030-45316-9_5

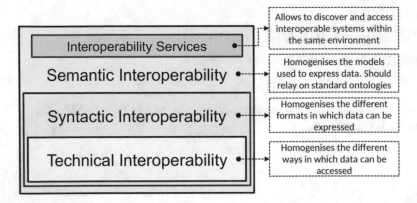

Fig. 5.1 Semantic interoperability framework

the ontologies, which are the cornerstone of these technologies, allow to define terms unambiguously easing their reusability and allowing to infer knowledge thanks to the fact that the axioms defined by ontologies are described in the formal logic. As a result, a large number of ontologies have been built by both research and industry communities for addressing interoperability challenges. Furthermore, there are several standardisation bodies such as the W3C and the European Telecommunications Standards Institute (ETSI), which also support and promote ontologies to address the interoperability challenges.

The rest of this chapter is organised as follows: Section 2 describes how to develop ontologies for the semantic interoperability. Section 3 explains how the framework for the semantic interoperability is implemented, along with, the interoperability services for transparent discovery and access of IoT infrastructures. Section 4 contains conclusion that summarises the main topics of this chapter.

2 Ontologies

In computer science, ontologies are defined as "formal, explicit specifications of a shared conceptualisation" [25]. The ontologies are formal in the sense of description logics [2] they follow and are implemented in a machine-readable format following the W3C Web Ontology Language standard OWL [3].

Ontologies play an essential role for supporting the implementation of semantic interoperability, as they allow to explicitly describe and annotate the different components in the IoT domain, along with their functionalities and interactions. Agents, applications, or any information systems rely on the ontologies to specify what terms are involved in a data exchange and what these terms mean. The following section will show how to generate these ontologies focusing on the semantic interoperability.

2.1 Requirements for the Semantic Interoperability

When developing ontologies for the semantic interoperability there are some requirements that must be considered and fulfilled. In the following we give a list of such requirements:

- **Reuse of existing standardised ontologies.** Aiming at maximising the interoperability between different models it is recommended to reuse ontologies, or ontology terms, whenever it is possible. There are several catalogues that may help search for such ontologies by registering and indexing them, such as LOV [27], LOV4IoT [14], or the catalogue for smart cities [22].
- **Compliance with existing standardised ontologies.** When developing ontologies for maximising the interoperability between systems, it is also advisable to be compliant with existing standardised ontologies that already cover the domain or part of the domain to be described. Ensuring compliance with standardised ontologies allows to guarantee integrity and quality, since the existing ontologies follow specifications already approved by the standardisation bodies.
- **Compliance with existing standard specifications** It is recommended to be compliant with standard specifications, e.g., ISO norms [15], if they describe the area of concern. Even though they do not have associated ontologies, the description of concepts and interactions can be adopted by the ontologies to be generated.
- **Aim at general definitions.** In the scenario where the ontologies are created for the interoperability between systems, the definitions of concepts and properties tend to be general without including too restrictive logical constraints. Specialised concepts could be further defined in an extension of the ontology.
- **Orientation to web services.** It should be taken into account that in the IoT domain every concept of the physical world should have a virtual representation which would provide access to data by means of web services.

In addition to the general requirements proposed above, there are some other requirements related to increasing the readability of ontologies. This is a key feature to be considered when developing ontologies since the ontologies should be as reusable as possible:

- **Document the ontology.** The machine-readable code of an ontology should be accompanied by a human-oriented version of the ontology, such as an HTML file, in order to facilitate the understandability for humans.
- **Provide examples.** The HTML documentation of the ontology should also include some examples that show how to use the ontology, rather than including only the definitions of the concepts and relationships.
- **Include ontology metadata.** It is important to provide the ontology metadata within the code, in order to facilitate its reuse. Some examples for metadata that should be included in the ontology code are its version, the license that applies to the ontology, the publishing organisation, or a contact point.

- **Publish your ontology.** If an ontology is meant to be reused, it should be published on the Web following the best practices. The ontology implementation and documentation should be reachable when requesting the ontology URI by means of content negotiation mechanisms.

2.2 Developing Ontologies for Semantic Interoperability

To develop ontologies a systematic approach should be followed, in order to guarantee quality and to address the requirements listed above. There are some works that try to align the ontology development process with agile software developing methodologies and practices in order to ease the integration between ontologies and software. This section aims to describe the Linked Open Terms (LOT)[1] methodology, which is an ontology development methodology organised in sprints and iterations in order to align the ontology development with software agile practices. The LOT methodology was proposed in [20] and further developed within the EU project VICINITY [10].

Figure 5.2 shows an overview of the activities that have to be performed during the LOT development process and the products as results of these activities. This

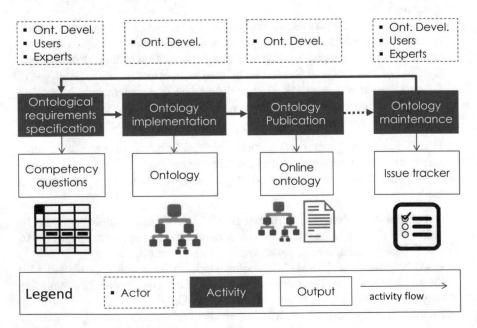

Fig. 5.2 Ontology development workflow with inputs, outputs, and actors

---

[1] http://lot.linkeddata.es.

development approach includes four activities, namely: the ontological requirements specification, the ontology implementation, the ontology publication, and the ontology maintenance.

2.2.1   Ontological Requirements Specification

The aim of the first activity, *the ontological requirements specification*, is to state why the ontology is being built and to identify and define the requirements the ontology should fulfil. In this activity, involvement and commitment of experts in the specific domain is required, in order to generate the appropriate industry perspective and knowledge.

To define such requirements, the necessary documentation about the domain to be modelled is needed, such as manuals, datasets, or standards. Taking the documentation and data provided by domain experts and users as input, the ontology development team generates a first proposal of ontological requirements written in the form of Competency Questions [12] or natural language sentences. In order to ease the management of the ontological requirements, it is proposed to store them by using a tabular approach including the following fields:

- Requirement identifier, which should be unique for each requirement
- Competency question or statement
- Provenance information
- Comments
- Relation with other requirements
- Priority
- Status, which can be proposed, accepted, rejected, or superseded by
- Sprint in which the requirements must be implemented

VICINITY Showcase Figure 5.3 shows one example where Google Spreadsheets are used to specify and store the ontological requirements. The spreadsheets were created by the ontology developers in the VICINITY project.

Identifier (domain+id)	Sprint	Competency Question / Natural language sentence (fact)	Answer	Requirement Status (Proposed, Accepted, Rejected, Deprecated)	Superseded by	Comments	Extracted from (provenance)
platform1	2	What is an organization?		A			D1.5 - Use Case 1 UC0100
platform2	2	What is an IoT device?		A			D1.5 - Use Case 1 UC0100
platform3		What is an add-value service?		R		Proposed to be deleted as the entity to be represented would be a more general service type.	D1.5 - Use Case 1 UC0100
platform4	2	What is a partnership?		A			D1.5 - Use Case 1 UC0100
platform5	2	What attributes has a partnership?		A			D1.5 - Use Case 1 UC0100

Fig. 5.3 Example of requirements in a Google Spreadsheet

Here you can find the list of the requirements identified for Vicinity core model ontology and their main features.

Identifier ⬍	Sprint ⬍	Competency Question ⬍	Answer ⬍	Extracted from ⬍	Priority ⬍
platform1	2	What is an organization?		D1.5 - Use Case 1 UC0100	
platform2	2	What is an IoT device?		D1.5 - Use Case 1 UC0100	
platform4	2	What is a partnership?		D1.5 - Use Case 1 UC0100	
platform5	2	What attributes has a partnership?		D1.5 - Use Case 1 UC0100	

Fig. 5.4 Example of requirements in an HTML document

These Google Spreadsheets are converted to an HTML file with the most relevant information for the users to facilitate visualisation. The VICINITY ontology portal[2] includes the requirements in HTML for each ontology. Figure 5.4 provides the example of the requirements for the VICINITY core model ontology.

2.2.2 Ontology Implementation

During the *ontology implementation* activity, the ontology is built using a formal language, based on the ontological requirements identified by the domain experts. After defining the first set of requirements, though modification and addition of requirements is allowed during the development, the ontology implementation phase is carried out through a number of sprints. The ontology developers schedule and

---

[2]http://vicinity.iot.linkeddata.es.

plan the ontology development according to the prioritisation of the requirements in *the ontology requirements specification* phase. The ontology development team builds the ontology iteratively, implementing only a certain number of requirements in each iteration.

The ontology implementation activity is divided into three sub-activities: conceptualisation, encoding, and evaluation. During the conceptualisation activity an ontology model is built from the ontological requirements identified in the requirements specification process. Then, during the encoding, the ontology development team generates computable models in the OWL language from the ontology model. Finally, the ontology developers should guarantee its quality and correctness by evaluating the ontology in different aspects:

- The ontology developers guarantee that the ontology does not have syntactic, modelling, or semantic errors. In this step reasoners or ontology validators such as OOPS! (OntOlogy Pitfall Scanner!) [23] could be used to make the process more simple.
- The ontology developers guarantee that the ontology fulfils the requirements scheduled for the ontology using the test cases generated in the requirements specification process. The tool Themis[3] could be used to verify that all the requirements are satisfied.
- The ontology developers verify that the ontology is compliant with the standards in the IoT domain in order to ensure integrity in the domain. The tool Themis could be also used to check whether the developed ontology addresses the requirements of some standards.

As there will be a new version of the ontology for each iteration, a versioning system should also be followed in order to avoid conflicts between versions. Therefore, the versioning identification will be as similar as possible to the conventions used in the software development. For example, each release could follow the pattern *v.major.minor.fix*, where each field follows these rules:

- **major**: This field is updated when the ontology covers the complete domain it intends to model. Thus, in this case, it is a complete product.
- **minor**: This field is updated when all the requirements of a subdomain are covered or when new documentation is added to the ontology.
- **fix**: This field is updated when typos or bugs are corrected in the ontology or when classes, relationships, axioms, individuals, or annotations are added, deleted, or modified.

VICINITY Showcase For addressing all the ontological requirements identified by the domain experts in the VICINITY project five ontologies were created, namely, the VICINITY Core (Core), the Web of Things (WoT), the WoT Mappings (Mappings), the VICINITY Adapters (Adapters), and the Datatypes (Datatypes) ontologies. All these ontologies are based on the standardised ontologies such as the

---

[3]http://themis.linkeddata.es.

SSN ontology,[4] the SOSA ontology,[5] and the SAREF4BLDG ontology [21], which are extended to fulfil the VICINITY requirements. How these ontologies look like and how they interact with one another can be seen in the VICINITY Ontology Portal.[6]

The Core ontology represents the information needed to exchange data between peers through an IoT platform; this ontology is being created by following a cross-domain approach and implements requirements from different domain experts. The WoT ontology aims at model the Web of Things domain according to the W3C WoT Working Group descriptions.[7] The Mappings ontology represents the mechanism for accessing the values provided by Web of Things in the VICINITY platform. The Adapters ontology aims to model all the different types of devices and properties that can be defined in the VICINITY ecosystem.

Finally, the Datatypes ontology aims to model the required and provided datatypes that are used in the interaction patterns, such as properties, actions, and events of the Things (smart devices and services) that are part of the IoT ecosystem such as the VICINITY.

Figure 5.5 shows an overview of the ontology network, where it can be seen how the ontologies interact with each other. In the figure it can also be observed how the VICINITY ontologies reuse terms from well-known standardised ontologies, such as the SSN ontology, the SOSA ontology, or the SAREF4BLDG ontology [21].

To make it visible and accessible online, each developed ontology can be stored in a GitHub repository, as it has been done in the VICINITY project.

Each GitHub repository includes a folder for each of the artifacts generated during the development process, e.g., the ontology code, the conceptualisation diagrams, or the ontological requirements.

Storing the developed ontologies in their corresponding GitHub repositories simplifies the ontology evaluation process.

The evaluation tool OnToology [1], which integrates OOPS! among other tools, generates automatically a new folder in a GitHub repository with all the evaluation report. Additionally, ontology developers also use Themis [9] to check whether all the requirements identified by the domain experts are satisfied by the ontology network.

In the VICINITY project Themis was also used to determine whether the VICINITY ontology network is compliant with several IoT standardised ontologies, namely, the SAREF ontology [7], the oneM2M ontology,[8] and the SSN ontology, and with other standards specifications, such as the ISO/IEC 30141:2017 [15] and

---

[4]http://www.w3.org/ns/ssn.

[5]http://www.w3.org/ns/sosa.

[6]http://vicinity.iot.linkeddata.es.

[7]https://www.w3.org/WoT/WG/.

[8]http://www.onem2m.org/technical/onem2m-ontologies.

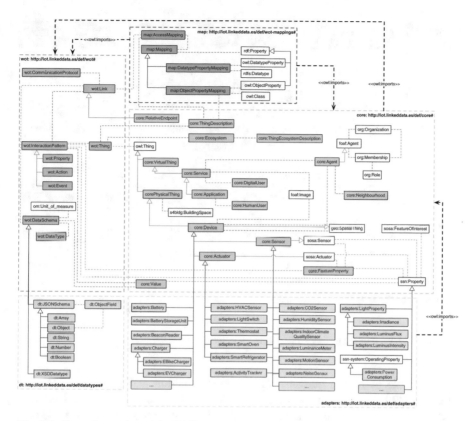

Fig. 5.5 General overview of the VICINITY ontology network

the OCF specification.[9] From this conformance analysis it was concluded that even though the ontology network does not address all the requirements identified for the standardised ontologies and standard specifications, there are not incompatibilities with them. All the information related to the conformance analysis is also openly available on the web site,[10] with the aim of allowing other developers to verify the results (Fig. 5.6).

2.2.3 Ontology Publication

The aim of the *ontology publication* activity is to publish the ontology following best practices for publishing vocabularies on the Web.[11] The ontologies should

---

[9]https://openconnectivity.org/.

[10]http://vicinity.iot.linkeddata.es/vicinity/conformance.html.

[11]https://www.w3.org/TR/swbp-vocab-pub/.

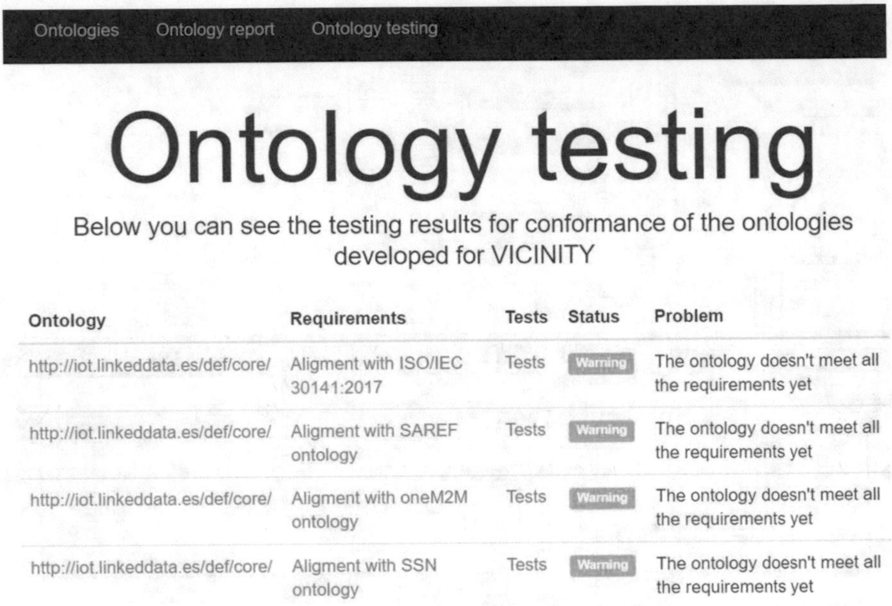

Ontologies   Ontology report   Ontology testing

Ontology testing

Below you can see the testing results for conformance of the ontologies developed for VICINITY

Ontology	Requirements	Tests	Status	Problem
http://iot.linkeddata.es/def/core/	Aligment with ISO/IEC 30141:2017	Tests	Warning	The ontology doesn't meet all the requirements yet
http://iot.linkeddata.es/def/core/	Aligment with SAREF ontology	Tests	Warning	The ontology doesn't meet all the requirements yet
http://iot.linkeddata.es/def/core/	Aligment with oneM2M ontology	Tests	Warning	The ontology doesn't meet all the requirements yet
http://iot.linkeddata.es/def/core/	Aligment with SSN ontology	Tests	Warning	The ontology doesn't meet all the requirements yet

Fig. 5.6 Overview of the testing section in the VICINITY ontology portal

be accessible both as a human-readable documentation and a machine-readable file from its URI. These two versions of the ontology, both the human-readable documentation and the machine-readable file, should be reached from the same URI using content negotiation mechanisms. There are tools that make this publication activity easier, such as OnToology.

It is worth noting that the machine-readable file is generated in the implementation activity, in which an OWL file is generated. Regarding the human-readable file, taking the ontology generated in the previous activity as the input, the ontology development team, in collaboration with the domain experts, generates the documentation of the release candidate. The documentation should describe the classes, properties, and data properties of the ontology, as well as the license URI and the title. The domain experts have to collaborate with the ontology development team to describe the classes and the properties. This description also includes metadata, such as creator, publisher, date of creation, last modification, or version number. The generation of this documentation could be automated using tools such as Widoco [11], which is also integrated in OnToology, to generate the HTML file with the metadata that is included in the ontology code.

VICINITY Showcase To support the ontology publication phase, the five VICINITY ontologies are published online to be accessible to everyone when looking up each ontology URI. In addition, the links to their published versions are also added

Fig. 5.7   Overview of VICINITY ontology portal

on the online VICINITY ontologies portal[12] to make them accessible to all the users from a single entry point.

Figure 5.7 shows an overview of the VICINITY ontology portal which was designed with the aim of providing the users with all the information of each ontology. As it can be observed from the figure, the portal provides the following information for each ontology: (a) a link to the ontology (code and documentation); (b) a brief description of the ontology domain; (c) a link to the repository where all the resources are managed; (d) a link to the issue tracker where new issues, bugs, and suggestions could be reported by any user; (e) a link to the ontology requirements; and (f) a link to the different ontology releases. Finally, the portal includes the top menu functionalities related to ontology testing, such as browsing alignments with standards, accessing ontology requirements tests, and testing the ontology network using Themis.

The HTML documentation for each of the implemented ontologies is automatically generated by the tool OnToology.

2.2.4   Ontology Maintenance

Last but not least, the goal of the last activity in the ontology development process is to update and add new requirements to the ontology that are not identified in

---

[12]http://vicinity.iot.linkeddata.es/vicinity/.

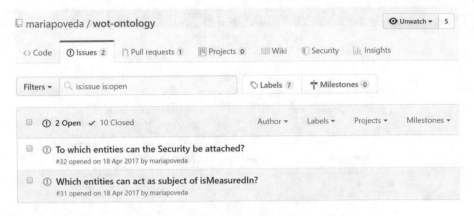

Fig. 5.8 GitHub issues related to the VICINITY core ontology

the existing set of requirements, to identify and correct errors or to schedule a new iteration for the ontology development. During the ontology development process, the domain experts can propose new requirements or improvements for the ontology. If these requirements or improvements are approved by the ontology development team, they are added to the ontology.

VICINITY Showcase To support the maintenance of the ontology, considering that the VICINITY ontology network follows an iterative development approach, the ontology developers use the GitHub issue tracker, shown in Fig. 5.8, which manages and maintains the list of issues identified by the domain experts and the ontology developers.

All the changes and improvements over the ontology need to be agreed by the whole ontology development team. The GitHub issue tracker is used to discuss improvements and issues about the domains. If the domain experts or the ontology developers want to add new concepts to the ontology they have to create a new issue in the GitHub issue tracker, which will be used to start a discussion about the approval of the proposal. GitHub issues will be marked as "open" and "closed". The issues can also have an assignee which is the person that is responsible for moving the issue forward.

The ontology developers should label each issue according to its topic or status in order to organise the different types of issues. GitHub comes with a few labels by default: "bug", "duplicate", "enhancement", "invalid", "question", and "won't fix". In addition, the ontology developers can create new labels if they think they can improve issue management. The ontology developers are also responsible for closing the issues created that have already been addressed.

3 Semantic Interoperability

The semantic interoperability can be viewed as a normalisation process of the data that addresses how data is accessed, the syntax of such data, and how data is modelled. The semantic interoperability consists of several layers that must be implemented in order to build an interoperable environment [17]. The technical interoperability layer, that defines how data is accessed, is commonly implemented by means of REST APIs, so IoT infrastructures can be managed as Web resources [4, 13]; although other alternatives exist [8]. The syntactic interoperability layer, that defines the syntax of data, is commonly addressed following two approaches [26]. The former consists in expressing data in RDF and leaving the content negotiation mechanisms to decide the specific format required by a data consumer. The latter agrees on using JSON or XML as formats to represent data since these two data formats are the most used by REST APIs, and rely on translation algorithms to convert such data into its respective RDF representation.

Once the technical and the syntactic interoperability layers are implemented, the semantic interoperability layer consists in deciding which ontologies will be used to model the data [19]. As a good practice the selected ontologies should follow the standards and, in the case the standardised ontology does not cover all the data requirements for a specific domain problem, it should be extended to cover these requirements [18]. Nevertheless, the core of the ontology will be always compliant with the standard.

Note that implementing the syntactic interoperability by just using a JSON or XML format is not enough, RDF should be the language used to express the data. There are two reasons for this. The first reason is that it allows to express data in different formats using content negotiation mechanisms, such as JSON–LD or RDF/XML. And the second reason is that it allows to model data using standard data models which in addition are formally defined, i.e., ontologies, which is not possible if just JSON or XML is used. In addition, the fact of using standards immediately enables any system to be understood by others using the same standard. Finally, another benefit of using RDF and ontologies is that they support the use of SPARQL language, which allows querying data expressed in RDF following a certain ontology.

VICINITY Showcase Figure 5.9 shows one example of the implementation of the semantic interoperability based on the implementation in the VICINITY project. In the VICINITY project the technical interoperability has been addressed by specifying a REST API. The new infrastructures to be integrated in the VICINITY cloud must translate their APIs into the VICINITY standard REST API by means of a component named "Adapter". Therefore, in the VICINITY the technical interoperability is implemented, and solved, thanks to this component. Once the adapters translates any API to the one defined in the VICINITY, the platform relies on another component, i.e., the "Gateway API", that publishes the data in a peer-to-peer network using its credentials and ensuring several security policies.

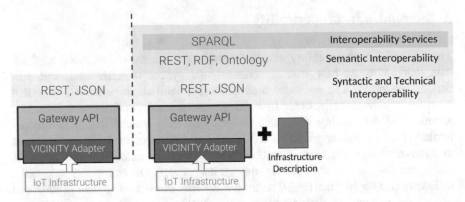

Fig. 5.9 VICINITY example: semantic interoperability implementation

The syntactic interoperability has been addressed by agreeing on the format the endpoints use, i.e., JSON format. However, no restriction on the set of key-values that these JSON documents may contain has been formulated.

The semantic interoperability implemented in the VICINITY is based on several standard ontologies such as the Web of Things (WoT) from the W3C and SAREF from the ETSI. It should be mentioned that the VICINITY version of the WoT ontology was developed based on the W3C requirements of the official WoT ontology, in fact, the VICINITY WoT model was taken as a seed model for the on-going W3C WoT ontology implementation. In addition, since the syntactic interoperability implemented does not rely on the use of RDF format but JSON, the ontology called WoT Mappings was developed to allow the on the fly translation of JSON data into RDF.

Finally, since all the data can be translated into RDF format following the ontologies all the Gateway APIs come with a query interface that implements SPARQL 1.1; SPARQL 1.1 is the W3C standard language to query RDF data [6].

3.1 Semantic Interoperability Services

The interoperability services are implemented inside the semantic interoperability layer. These services allow to discover IoT infrastructures in an interoperable environment, and then, access their data distributively [5]. The implementation of the semantic interoperability services requires three main steps: *register*, *discover*, and *distributed access*. The register phase consists in storing the description of an IoT infrastructure, the discovery consists in finding suitable descriptions to answer a given query, and finally, the distributed access consists in retrieving the relevant data from remote endpoints that a query may require in order to provide a result.

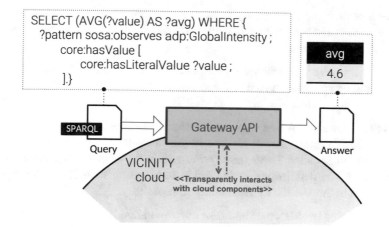

Fig. 5.10 VICINITY example: semantic interoperability services

VICINITY Showcase: Using the Semantic Interoperability Services For example, as shown in Fig. 5.10, a user may submit a query to its Gateway API asking for the average solar radiation detected by a set of sensors and receive an answer "4.6". The user does not need to be aware of the number of Gateway APIs that must be involved to answer the query, as well as, how they are accessed, their formats, and their models. The user submits a query, which is answered transparently by discovering relevant Gateway APIs, fetching their data, and aggregating the values within their data.

3.2 Register

Registration is the first step towards the implementation of an interoperable environment. When a new IoT infrastructure is included in the interoperable environment it has to register the description of its things such as devices it is composed of. The description usually contains:

- some metadata about the infrastructure, such as what kind of devices it is composed of,
- some contextual data such as city, or part of the building in which a device is located,
- a specification of how to translate its data into RDF if required, known as mapping, and
- a description of how to access the data provided by the devices of the IoT infrastructure

- a description if the devices have properties and how to access them, if they have some actions and how to interact with them, and finally, if they trigger events and how to listen to them.

These descriptions are paramount to enable the discovery and the distributed access. They should be specified in RDF, so they could be queryable, and should follow a standard ontology when possible.

VICINITY Showcase A user who wants his/her IoT infrastructure to become a part of the VICINITY must provide its description in JSON format. In the VICINITY this description is called *Thing Description* and in the rest of this section we will use this term. The detailed instruction guide for creating such *Thing Descriptions* can be found in the VICINITY Agent Github repository.[13]

For conversion of JSON data into their corresponding RDFs the created mapping ontology WoT Mappings[14] is used. Figure 5.11 shows an RDF example of an IoT infrastructure that is a part of the VICINITY platform. This simple IoT infrastructure consists of one weather station in the city of Martim Longo in Portugal, which is reporting the outdoor solar radiation.

The RDF contains some attributes such as the name of the infrastructure and the name of its VICINITY adapter. Note that the type of the IoT infrastructure, i.e., a weather station is also included. The type is defined in the VICINITY Adapters ontology.[15]

Figure 5.12 shows an RDF that includes description of the location of the IoT infrastructure. In this example, the infrastructure is located in a building called "Solar Lab", in a city named Martim Longo, in the country of Portugal.

VICINITY example: defining an IoT infrastructure

```
@prefix core: <http://iot.linkeddata.es/def/core#> .
@prefix adp: <http://iot.linkeddata.es/def/adapters#> .
@prefix wot: <http://iot.linkeddata.es/def/wot#> .
@prefix rdf: <http://www.w3.org/1999/02/22-rdf-syntax-ns#> .

<http://vicinity.eu/data/things/fd5b6c22> a wot:Thing;
  wot:name "DATATAKER-sll";
  wot:adapter-id "DATATAKER-sll";
  core:represents [
    rdf:type adp:WeatherStation;
  ].
```

Fig. 5.11 VICINITY example: defining an IoT infrastructure

---

[13] https://github.com/vicinityh2020/vicinity-agent.

[14] http://iot.linkeddata.es/def/wot-mappings/index-en.html.

[15] http://iot.linkeddata.es/def/adapters.

```
VICINITY example: defining the location of an IoT infrastructure
```

```
@prefix core: <http://iot.linkeddata.es/def/core#> .
@prefix rdfs: <http://www.w3.org/2000/01/rdf-schema#> .
@prefix saref4city: <https://w3id.org/def/saref4city#> .
@prefix rdf: <http://www.w3.org/1999/02/22-rdf-syntax-ns#> .
@prefix s4bldg: <https://w3id.org/def/saref4bldg#> .

<http://vicinity.eu/data/things/fd5b6c22-301d-4697-8d5f-3e7a472d21df
core:isLocatedAt [
    rdf:type s4bldg:Building ;
    rdfs:label "Solar Lab";
  ], [
    rdf:type saref4city:City ;
    rdfs:label "Martim Longo";
  ], [
    rdf:type saref4city:Country ;
    rdfs:label "Portugal" ;
  ] .
```

Fig. 5.12 VICINITY example: defining the location of an IoT infrastructure

For describing location the classes defined in the ontologies SARE4CITY[16] and SAREF4BLD[17] are used. These ontologies are extension of the standard SAREF ontology[18] for city and building domains, respectively.

The description may also specify who is the owner of a certain infrastructure as it is shown in Fig. 5.13. In this example, the owner has a *name* "Enercoutim" and an *openid* "7cc54178".

Finally, the description may contain the interaction patterns that devices of an infrastructure provide. This part of the description follows the VICINITY extended version of the Web of Things W3C specification,[19] and thus, three kind of patterns may be specified: properties, actions, and events.

Figure 5.14 specifies that the weather station of the IoT infrastructure described by Fig. 5.11 has an interaction pattern that is a property. The property has a name "Outdoor solar radiation" and measures the global solar radiation. In addition, the link from where the data of this property can be retrieved (see field "wot:href") and that this interaction pattern has a value are added. Note that the value is described by an URI that is http://vicinity.eu/data/descriptions/2cab232c; such URI is the identifier of the aforementioned WoT Mappings ontology that allows the on the

---

[16]https://w3id.org/def/saref4city.

[17]https://w3id.org/def/saref4bldg.

[18]https://ontology.tno.nl/saref/.

[19]http://iot.linkeddata.es/def/wot.

VICINITY example: defining the owner of an IoT infrastructure

```
@prefix core: <http://iot.linkeddata.es/def/core#> .
@prefix foaf: <http://xmlns.com/foaf/0.1/>
@prefix rdf: <http://www.w3.org/1999/02/22-rdf-syntax-ns#>

<http://vicinity.eu/data/things/fd5b6c22> core:hasOwner
 <http://vicinity.eu/data/things/7cc54178>;
<http://vicinity.eu/data/things/7cc54178> rdf:type core:Agent;
  foaf:name "Enercoutim";
  foaf:openid "7cc54178" .
```

Fig. 5.13 VICINITY example: defining the owner of an IoT infrastructure

VICINITY example: defining the interaction patterns of an IoT infrastructure

```
@prefix core: <http://iot.linkeddata.es/def/core#> .
@prefix sosa: <http://www.w3.org/ns/sosa/> .
@prefix wot: <http://iot.linkeddata.es/def/wot#> .
@prefix adp: <http://iot.linkeddata.es/def/adapters#> .
@prefix rdf: <http://www.w3.org/1999/02/22-rdf-syntax-ns#>

  <http://vicinity.eu/data/things/fd5b6c22> wot:providesInteractionPattern [
      rdf:type wot:Property .
      wot:interactionName "Outdoor solar radiation";
      sosa:observes adp:GlobalSolarRadiation;
      wot:isReadableThrough [
        wot:href "/objects/fd5b6c22/properties/GlobalSolarRadiation";
        wot:outputRawJSONString
        "{
          "type": "object",
          "field": [
            {
              "schema": {
              "type": "double"
              },
             "predicate": "core:value",
             "name": "GlobalSolarRadiation"
            };
          ]
        }";
      core:hasValue [
        a core:Value;
        core:describedBy <http://vicinity.eu/data/descriptions/2cab232c>;
      ];
    ].
```

Fig. 5.14 VICINITY example: defining the interaction patterns of an IoT infrastructure

VICINITY example: defining the translation mapping related to an IoT infrastructure

```
@prefix core: <http://iot.linkeddata.es/def/core#> .
@prefix map: <http://iot.linkeddata.es/def/wot-mappings#> .
@prefix wot: <http://iot.linkeddata.es/def/wot#> .
@prefix rdf: <http://www.w3.org/1999/02/22-rdf-syntax-ns#>

<http://vicinity.eu/data/descriptions/2cab232c> rdf:type core:ThingDescription;
  core:describes <http://vicinity.eu/data/things/2cab232c>;
  core:identifier "2cab232c";
  map:hasAccessMapping [
    rdf:type map:AccessMapping;
    map:hasMapping [
      rdf:type map:Mapping;
      map:key "observed-value";
      map:predicate core:literalValue ;
    ];
    map:mapsResourcesFrom [
      rdf:type wot:Link;
      wot:hasMediaType "application/json";
      wot:href "/objects/fd5b6c22/properties/GlobalSolarRadiation"
    ];
  ].
```

Fig. 5.15 VICINITY example: defining the translation mapping related to an IoT infrastructure

fly translation of the JSON document containing the value of the property to the RDF. The JSON document is specified in the field "wot:outputRawJSONString".

The defined mapping is shown in the example depicted by Fig. 5.15. Relying on this mapping the interoperability services are able to fetch the data exposed by this IoT infrastructure and translate it into RDF. The mapping is mainly composed of two parts, the former specifies how data is translated from JSON to RDF, and the latter specifies from where the data is fetched.

The mapping in Fig. 5.15 expects to receive a JSON document with a key that is "GlobalSolarRadiation". As stated above, the JSON is specified in the field "wot:outputRawJSONString" of the RDF in Fig. 5.14. Assuming that the fetched value for that key is "4.6", then, the mapping generates the RDF depicted in Fig. 5.16. Note that the value of the interaction pattern in the translated RDF is fetched using the href endpoint specified in the mapping field "wot:href" in Fig. 5.15.

In the VICINITY the translated RDFs of all the Thing Descriptions are stored in a centralised component known as Semantic Repository. Figure 5.17 summarises the registration process implemented in the VICINITY. A user registers a Thing Description of his/her infrastructure in JSON. Then this JSON is automatically translated to RDF and sent to the Semantic Repository cloud component, where it is stored to enable the discovery of the IoT infrastructure by third-party Gateway APIs that are granted to discover that infrastructure. Discovery permissions are granted by the user.

VICINITY example: RDF translated from JSON by means of a VICINITY mapping

```
@prefix core: <http://iot.linkeddata.es/def/core#> .
@prefix wot: <http://iot.linkeddata.es/def/wot#> .

<http://vicinity.eu/data/things/fd5b6c22> wot:providesInteractionPattern [
    core:hasValue [
        a core:Value;
        core:literalValue "4.6";
    ];
    ].
```

Fig. 5.16 VICINITY example: RDF translated from JSON by means of a VICINITY mapping

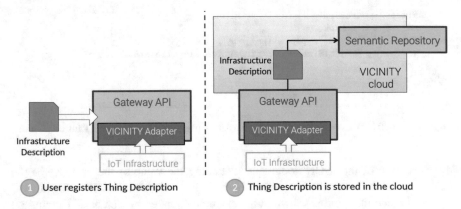

Fig. 5.17 VICINITY example: registering a thing description of an IoT infrastructure

3.3 Discovery

The discovery within the semantic interoperability services allows to find suitable IoT infrastructures that fulfil the requirements of a given query. As a requirement, the discovery needs to have a set of their descriptions over which the discovery process takes place. We distinguish two types of queries:

- *discovery queries.* Queries that refer only to information that is provided in the description of the registered IoT infrastructure.
- *access queries.* Queries that require to access the infrastructures and fetch data produced by their devices in order to answer the query.

For both types of queries discovery is the first step that has to be performed to allow queries.

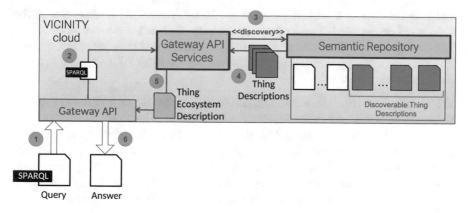

Fig. 5.18 VICINITY example: semantic interoperability services, discovery

VICINITY Showcase Figure 5.18 shows one example how the discovery process can be implemented using the standard query language SPARQL that is the part of the W3C standards of the semantic web. The example demonstrates the discovery process implemented in the VICINITY project. In the VICINITY the discovery is performed by the cloud component called Gateway API Services[20] which receives a SPARQL query and a set of security parameters from a Gateway API, and returns a set of descriptions of devices in IoT infrastructures that meet the requirements of the query. In the case of *discovery query* only, the Gateway API Services interact with the semantic repository and find the relevant *Thing Descriptions* that the Gateway API that requested the query is allowed to access. Then, the Gateway API Services aggregate the Thing Descriptions into a document called Thing Ecosystem Description, which is a set of Thing Descriptions, and send it back to the Gateway API. The Gateway API receives the Thing Ecosystem Description via its Query Distributed Client. Then, the client computes a query answer using information contained in the received Thing Ecosystem Description.

In the case of *access query* the client first checks if some devices must be accessed by the Gateway API before computing and providing an answer to the query. The access of devices is done within the step *distributed access* described in the following.

3.4 Distributed Access

The distributed access within the semantic interoperability services is triggered when a query requires also to include some remote data in order to be answered,

---

[20]https://github.com/vicinityh2020/vicinity-gateway-api-services.

and thus the data of a remote endpoint has to be fetched. As a requirement, the distributed access needs first to perform the step *discovery* described in Sect. 3.3. The description of IoT infrastructures must be complete and include the information about the remote endpoints of their corresponding Gateway APIs in order to allow their automatic access.

VICINITY Showcase: Accessing Infrastructures Similar to a *discovery query*, an *access query* is sent to the Gateway API Services. This component computes a Thing Ecosystem Description with the relevant Thing Descriptions to answer a given query. However, this time the Gateway API Services detect that the query needs to access remote Gateway APIs and write this fact in the Thing Ecosystem Description. Then, when the Distributed Query Client receives the Thing Ecosystem Description, it detects that the data of some Gateway APIs must be fetched.

In the second step the Gateway API accesses the remote endpoints of the other Gateway APIs in its peer-to-peer network and provides the JSON documents fetched to the Distributed Query Client. This latter component translates the JSON documents into RDF using the mappings within the Thing Ecosystem Description, and finally, includes the RDF translated into the Thing Ecosystem Description. Figure 5.19 depicts the distributed access process aforementioned.

Finally, the query is answered and an answer to this query is provided. Figure 5.20 reports how the fetched data from the different Gateway APIs is translated into RDF in order to answer the query.

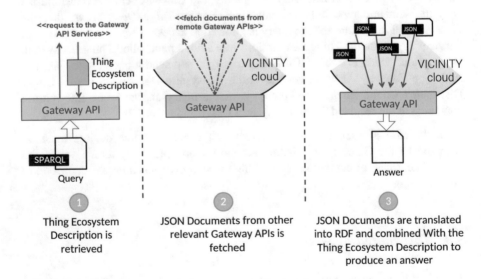

Fig. 5.19 VICINITY example: semantic interoperability services, accessing

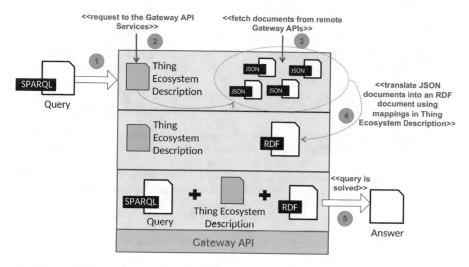

Fig. 5.20 VICINITY example: semantic interoperability services, query solving

4   Conclusions

The semantic interoperability enables all the systems in an environment to communicate transparently. Due to the nature of the IoT infrastructures, which publish data but also may interact with one another, the interoperability has become a cornerstone component to count with in the IoT context.

The interoperability among IoT infrastructures requires a normalisation of the data access methods (technical interoperability layer), the format used to represent data (syntactic interoperability layer), and the form under which data is model (semantic interoperability layer).

The implementation of the interoperability layers has been briefly described by relying on the semantic technologies. The technical layer has been solved by publishing data by means of REST APIs on the Web, the syntactic by using RDF either in the published data or translating other formats on the fly, and the semantic by relying on ontologies.

The ontologies are the pillar component when implementing an interoperable environment, and specifically to implement the semantic interoperability. The ontologies allow to define terms and relationships in a problem domain without ambiguity, and allow to establish a consensus between different communities about the meaning and the use of data.

As a result of implementing the different interoperability layers, the implementation of the interoperability services on top of such layers is feasible. These services allow transparent discovery of and access to IoT infrastructures in order to answer a given query.

References

1. Alobaid, A., Garijo, D., Poveda-Villalón, M., Santana-Perez, I., Fernández-Izquierdo, A., & Corcho, O. (2019). Automating ontology engineering support activities with OnToology. *Journal of Web Semantics, 57*, 100472.
2. Baader, F., Calvanese, D., McGuinness, D., Patel-Schneider, P., & Nardi, D. (2003). *The description logic handbook: Theory, implementation and applications.* Cambridge: Cambridge University Press (2003)
3. Bechhofer, S., Van Harmelen, F., Hendler, J., Horrocks, I., McGuinness, D. L., Patel-Schneider, P. F., Stein, L. A., et al. (2004). OWL web ontology language reference. *W3C Recommendation, 10*(2)
4. Botta, A., De Donato, W., Persico, V., & Pescapé, A. (2016). Integration of cloud computing and Internet of Things: A survey. *Future Generation Computer Systems, 56*, 684–700.
5. Cimmino, A., Poveda-Villalón, M., & García-Castro, R.: VICINITY: IoT semantic interoperability based on the web of things. In *International Workshop on Security and Reliability of IoT Systems.*
6. Harris, S., Seaborne, A., & Prud'hommeaux, E. (2013). SPARQL 1.1 query language. *W3C Recommendation, 21*(10), 778.
7. Daniele, L., den Hartog, F., & Roes, J. (2015). Created in close interaction with the industry: The smart appliances reference (SAREF) ontology. In *International Workshop Formal Ontologies Meet Industries* (pp. 100–112). Berlin: Springer.
8. Derhamy, H., Eliasson, J., Delsing, J., & Priller, P. (2015). A survey of commercial frameworks for the Internet of Things. In *2015 IEEE 20th Conference on Emerging Technologies & Factory Automation (ETFA)* (pp. 1–8). Piscataway: IEEE.
9. Fernández-Izquierdo, Alba, García-Castro, & Raúl (2019). Themis: a tool for validating ontologies through requirements. *Proceedings of the International Conference on Software Engineering and Knowledge Engineering, SEKE, 2019*, KSI Research Inc. and Knowledge Systems I, 573–578.
10. García-Castro, R., Fernández-Izquierdo, A., Heinz, C., Kostelnik, P., Poveda-Villalón, M., & Serena, F. (2017). D2.2 detailed specification of the semantic model. Tech. Rep., Universidad Politécnica de Madrid (UPM). VICINITY Project. https://vicinity2020.eu
11. Garijo, D. (2017). WIDOCO: A wizard for documenting ontologies. In *International Semantic Web Conference* (pp. 94–102). Berlin: Springer.
12. Grüninger, M., & Fox, M. S. (1995). Methodology for the design and evaluation of ontologies. *IJCAI'95, Workshop on Basic Ontological Issues in Knowledge Sharing.*
13. Guinard, D., Trifa, V., Mattern, F., & Wilde, E. (2011). From the Internet of Things to the web of things: Resource-oriented architecture and best practices. In *Architecting the Internet of Things* (pp. 97–129). Berlin: Springer.
14. Gyrard, A., Atemezing, G., Bonnet, C., Boudaoud, K., & Serrano, M. (2016). Reusing and unifying background knowledge for internet of things with LOV4IoT. In: *2016 IEEE 4th International Conference on Future Internet of Things and Cloud (FiCloud)* (pp. 262–269). Piscataway: IEEE.
15. International Organization for Standardization Geneva, S. (2017). ISO/IEC 30141:2017: Internet of Things (IoT) - reference architectures. https://www.iso.org/standard/65695.html
16. Kosek, A. M., Syed, A. A., & Kerridgey, J. M. (2010). RDF recipes for context-aware interoperability in pervasive systems. In *The IEEE Symposium on Computers and Communications* (pp. 1017–1022). Piscataway: IEEE.
17. Noura, M., Atiquzzaman, M., & Gaedke, M. (2018). Interoperability in Internet of Things: Taxonomies and open challenges. *Mobile Networks and Applications, 24*, 796–809.
18. Noy, N., McGuinness, D., & Hayes, P. J. (2005). Semantic Integration & Interoperability Using RDF and OWL.

19. Obrst, L. (2003). Ontologies for semantically interoperable systems. In *Proceedings of the Twelfth International Conference on Information and Knowledge Management* (pp. 366–369). New York: ACM.
20. Poveda-Villalón, M. (2012). A reuse-based lightweight method for developing linked data ontologies and vocabularies. In *Extended Semantic Web Conference* (pp. 833–837). Berlin: Springer (2012)
21. Poveda-Villalón, M., & García-Castro, R. (2018). Extending the SAREF ontology for building devices and topology. In *Proceedings of the 6th Linked Data in Architecture and Construction Workshop (LDAC 2018), CEUR-WS* (vol. 2159, pp. 16–23).
22. Poveda-Villalón, M., García-Castro, R., & Gómez-Pérez, A. (2014). Building an ontology catalogue for smart cities. In *eWork and eBusiness in Architecture, Engineering and Construction: ECPPM 2014* (vol. 1, pp. 829–839). Leiden: CRC Press. http://oa.upm.es/36715/
23. Poveda-Villalón, M., Gómez-Pérez, A., & Suárez-Figueroa, M. C. (2014). Oops!(ontology pitfall scanner!): An on-line tool for ontology evaluation. *International Journal on Semantic Web and Information Systems, 10*(2), 7–34.
24. Serena, F., Poveda-Villalón, M., & García-Castro, R. (2017). Semantic discovery in the web of things. In *International Conference on Web Engineering* (pp. 19–31). Berlin: Springer.
25. Studer, R., Benjamins, V. R., & Fensel, D. (1998). Knowledge engineering: Principles and methods. *Data & Knowledge Engineering, 25*(1–2), 161–197.
26. Toma, I., Simperl, E., & Hench, G. (2009). A joint roadmap for semantic technologies and the internet of things. In *Proceedings of the Third STI Roadmapping Workshop, Crete* (vol. 1, pp. 140–53).
27. Vandenbussche, P. Y., Atemezing, G. A., Poveda-Villalón, M., & Vatant, B. (2017). Linked open vocabularies (LOV): A gateway to reusable semantic vocabularies on the Web. *Semantic Web, 8*(3), 437–452.

Chapter 6
Standards for the IoT

Keith Dickerson, Raúl García-Castro, Peter Kostelnik, and Marek Paralič

1 Introduction

The IoT ecosystem is incredibly large and complex with many different interfaces and protocols and many competing products and services from different vendors which must all work together. Therefore, standards are important primarily because they enable interoperability between IoT systems, subsystems, devices, and applications.

> Machines and devices using new technologies create great opportunities for businesses to provide new services. However, these machines and devices need to be interoperable, or there will be barriers to cross-border business. Such interoperability will also allow businesses to mix and change suppliers and thus have more choice. Standardisation in key areas would greatly help this interoperability, without reducing innovation.[1]

Although standards in key areas are essential, especially in the context of the IoT-enabled digital single market quoted above, the temptation to create new standards where gaps are identified must be resisted unless absolutely necessary. The aim

---

[1] https://ec.europa.eu/digital-single-market/en.

K. Dickerson
Climate Associates Ltd, Ipswich, UK
e-mail: keith.dickerson@mac.com

R. García-Castro (✉)
Universidad Politécnica de Madrid, Madrid, Spain
e-mail: rgarcia@fi.upm.es

P. Kostelnik · M. Paralič
Intersoft A.S., Kosice, Slovakia
e-mail: peter.kostelnik@intersoft.sk; marek.paralic@intersoft.sk

© Springer Nature Switzerland AG 2021
C. Zivkovic et al. (eds.), *IoT Platforms, Use Cases, Privacy, and Business Models*,
https://doi.org/10.1007/978-3-030-45316-9_6

should always be to re-use existing standards wherever possible and to propose extensions or modifications to these if it is not possible to use them as they are.

The Standards groups, fora, and consortia relevant to the IoT are identified in Sect. 3. There are a large number of groups ranging from formal standards bodies such as ISO/IEC and ITU, which develop standards recognized internationally, through to fora and consortia which develop industry specifications and agreements which may be used as "de-facto" standards or promoted as international standards at a later date. The IoT Standards ecosystem is continually evolving with new groups being formed to fill perceived gaps, while other groups merge in response to changing marketplaces and requirements. In the future, standards will be key to providing the interoperability required between platforms, devices, gateways, servers, and applications for the development of cross-domain use cases to provide many benefits in environments such as Smart Cities. Without effective standards at all levels, the deployment of use cases such as these will not be possible.

2 Standard-Based Ontologies for the IoT

Current IoT deployments use a multi-tiered architecture that combines nodes, hubs, and cloud-based services. In the absence of a common systems engineering approach to specify where different types of rules are applied, nodes cannot know where the data they supply will be interpreted, or even that it will be interpreted only once. Contextualizing the data they send will increase the likelihood that it can be interpreted correctly. That contextualization should reference the most primitive possible schema or data model that results in a correct understanding, in order to increase further the chance of correct interpretation and to avoid leakage of unnecessary data about the system to observers. IoT interoperability should be supported by employing a generic IoT ontology based on existing standards such as W3C, ETSI, and oneM2M, to interchange IoT data in a range of standardized and proprietary formats. This support for interoperability will be extended to address the specific requirements of cross-IoT-domain value added services. Key challenges are:

- Ontologies that formalize the meaning of domain data and information models
- Ontology merging, matching, and alignment strategies across domains
- Semantic metadata
- Solving technical interoperability, which is usually associated with hardware/-software components, systems, and platforms that enable machine-to-machine communication to take place. This kind of interoperability is centred on (communication) protocols and the infrastructure needed for those protocols to operate.
- The messages transferred by communication protocols need to have a well-defined syntax and encoding, even if it is only in the form of bit-tables. However, many protocols carry data or content, and this can be represented using high-level formats.

- Semantic Interoperability: is usually associated with the meaning of content and concerns the human rather than machine interpretation of the content. Thus, interoperability on this level means that there is a common understanding between people of the meaning of the content (information) being exchanged.

2.1 W3C/OGC Semantic Sensor Network Ontology

OWL-S is an ontology built using the Web Ontology Language (OWL). The Semantic Sensor Network ontology (commonly known as "SSN") is an OWL-2 DL ontology for describing sensors and the observations they make of the physical world.[2] SSN is based on the Open Geospatial Consortium (OGC) Sensor Web Enablement standards (SensorML and Observations and Measurements) and is published in a modular architecture that supports the judicious use of "just enough" ontology for diverse applications, including satellite imagery, large scale scientific monitoring, industrial and household infrastructure, citizen observers, and Web of Things (WoT).

The SSN ontology contains concepts and relations relevant only to sensors, actuators, and sampling; leaving concepts related to other, or multiple, domains to be included from other ontologies when the ontology is used. Doing so makes the ontology single subject and so addresses modularity and reusability. The ontology describes sensors, the accuracy, etc., of such sensors, observations, and methods used for sensing. In addition, concepts for operating and survival ranges are included, as these are often part of a given specification for a sensor, along with its performance within those ranges. Finally, a structure for field deployments is included to describe deployment lifetime and sensing purpose of the deployed macro instrument. Modelling of concepts such as units of measurement, locations, hierarchies of sensor types, and feature and property hierarchies are left to other ontologies. The intention was to create core sensor description ontology, which can be easily extended with specific domain concepts.

2.2 OGC SensorML: Sensor Model Language

The primary focus of the Sensor Model Language (SensorML) is to provide a robust and semantically tied means of defining processes and processing components associated with the measurement and post-measurement transformation of observations.[3] This includes sensors and actuators as well as computational processes applied pre- and post-measurement. The main objective is to enable

---

[2]https://www.w3.org/TR/vocab-ssn/.

[3]SensorML Specification, Version 1.0.1., http://www.ogcnetwork.net/SensorML_Spec.

interoperability, first at the syntactic level and later at the semantic level (by using ontologies and semantic mediation), so that sensors and processes can be better understood by machines, utilized automatically in complex workflows, and easily shared between intelligent sensor web nodes. This standard is one of several implementation standards produced under OGC's Sensor Web Enablement (SWE) activity. This standard is a revision of content that was previously integrated in the SensorML version 1.0 standard (OGC 07-000).

SensorML is a means by which sensor systems or processes can make themselves known and discoverable. SensorML provides a rich collection of metadata that can be mined and used for discovery of sensor systems and observation processes. This metadata includes identifiers, classifiers, constraints (time, legal, and security), capabilities, characteristics, contacts, and references, in addition to inputs, outputs, parameters, and system location.

It can provide a complete and unambiguous description of the lineage of an observation, and it can describe in detail the process by which an observation happened. The original driver was to enable discovery of sensors distributed over the web, and to execute their services on-demand without a-priori knowledge of the sensor or processor characteristics. The self-describing characteristic of SensorML-enabled sensors and processes also supports the development of auto-configuring sensor networks, as well as the development of autonomous sensor networks in which sensors can publish alerts and tasks to which other sensors can subscribe and react. Finally, SensorML provides a mechanism for archiving fundamental parameters and assumptions regarding sensors and processes, so that observations from these systems can still be reprocessed and improved long after the origin mission has ended.

SensorML is encoded in XML Schema. However, the models and encoding pattern for SensorML follow Semantic Web concepts of Object-Association-Object. Therefore, SensorML models could easily be encoded for the Semantic Web. In addition, SensorML makes extensive use of soft-typing and linking to online dictionaries for definition of parameters and terms.

2.3 SenML: Sensor Markup Language

SenML defines media types for representing simple sensor measurements and device parameters in the Sensor Markup Language (SenML).[4] Representations are defined in JavaScript Object Notation (JSON), eXtensible Markup Language (XML), and Efficient XML Interchange (EXI), which share the common SenML data model. A simple sensor, such as a temperature sensor, could use this media type

---

[4]Media Types for Sensor Markup Language (SENML), https://tools.ietf.org/html/draft-jennings-senml-10.

in protocols such as HTTP or CoAP in order to either transport the measurements or to be configured.

SenML is designed so that processors with very limited capabilities could easily encode a sensor measurement into the media type, while at the same time a server parsing the data could relatively efficiently collect a large number of sensor measurements. There are many types of more complex measurements and measurements that this media type would not be suitable for. When developing the standard, a decision was made not to carry most of the metadata about the sensor in this media type to help reduce the size of the data and improve efficiency in decoding. The markup language can be used for a variety of data flow models, most notably data feeds pushed from a sensor to a collector, and the web resource model where the sensor is requested as a resource representation (GET /sensor/temperature). The main design goal is to be able to send simple sensor measurements in small packets on mesh networks from large numbers of constrained devices.

2.4   oneM2M Base Ontology

Ontologies are used in oneM2M (see Sect. 3.3) to provide syntactic and semantic interoperability of the oneM2M System with external systems.[5] These external systems are expected to be described by ontologies. The only ontology that is specified by oneM2M is the oneM2M Base Ontology formalized in OWL. The oneM2M Base Ontology is the minimal ontology that is required such that other ontologies can be mapped into oneM2M. The Base Ontology has been designed with the intent to provide a minimal number of concepts, relations, and restrictions that are necessary for semantic discovery of entities in the oneM2M System. To make such entities discoverable in the oneM2M System they need to be semantically described as classes (concepts) in a—technology/vendor/other-standard specific— ontology and these classes (concepts) need to be related to some classes of the Base Ontology as sub-classes. Additionally, the Base Ontology enables non-oneM2M technologies to build derived ontologies that describe the data model of the non-oneM2M technology for the purpose of interworking with the oneM2M System. The Base Ontology only contains Classes and Properties but not instances because the Base Ontology and derived ontologies are used in oneM2M to only provide a semantic description of the entities they contain. Instantiation (i.e., data of individual entities represented in the oneM2M System—e.g., devices, things, etc.) is done via oneM2M resources.

---

[5]oneM2M Base Ontology, Latest Draft Specifications, http://www.onem2m.org/technical/latest-drafts.

2.5 SAREF: Smart Applications REFerence Ontology

The Smart Appliances REFerence (SAREF) ontology is a shared model that facilitates the matching of existing assets (standards/protocols/datamodels/etc.) for developing smart applications.[6] The SAREF ontology provides building blocks that allow separation and recombination of different parts of the ontology depending on specific needs.

The starting point of SAREF is the concept of Device (e.g., a switch). Devices are tangible objects designed to accomplish one or more functions in households, common public buildings, or offices. The SAREF ontology offers a list of basic functions that can be eventually combined in order to have more complex functions in a single device. For example, a switch offers an actuating function of type "switching on/off". Each function has some associated commands, which can also be picked up as building blocks from a list. For example, the "switching on/off" is associated with the commands "switch on", "switch off", and "toggle". Depending on the function(s) it accomplishes, a device can be found in some corresponding states that are also listed as building blocks.

The concept Device offers a Service, which is a representation of a Function to a network that makes the function discoverable, registerable, and remotely controllable by other devices in the network. A Service can represent one or more functions. The Service is offered by a device that wants (a certain set of) its function(s) to be discoverable, registerable, remotely controllable by other devices in the network. The Service must specify the device that is offering the service, the function(s) to be represented, and the (input and output) parameters necessary to operate the service. The Device in the SAREF ontology is also characterized by an (Energy/Power) Profile that can be used to optimize the energy efficiency in a home or office that are part of a building.

2.6 From Ontologies of Things to Ontologies of Services

An interoperability solution aims to connect different isolated IoT infrastructures in order to create added value from them. However, such value creation depends on the effective collaboration between heterogeneous networks of devices and services. Note that in some cases the devices and services may be cross-domain such as involving health and buildings.

As discussed above, collaboration between different IoT infrastructures requires achieving semantic interoperability between them, so devices and data can be discovered, and data can be interchanged and understood among the different infrastructures.

---

[6]SAREF: the Smart Appliances REFerence ontology: https://saref.etsi.org/.

This requires, on the one hand, the enhancement of IoT data with metadata that describes its context (source, time, location, etc.) and, on the other hand, the representation of such data (and metadata) using ontologies that express the shared meaning of the data to ensure data consistency.

In order to ensure the common way of lookup and matching of devices and their services, the IoT ontology must contain a rich model of services to enable intelligent service discovery and execution. These concepts are based on the well-known service oriented architecture (SOA) approach. In SOA, distributed information systems enable loose coupling of system elements, i.e. various functional modules that provide and/or consume shared or private information resources, in a transparent way, by means of standardized service interfaces. The core concepts, which should be taken into account when designing the semantic service models are:

- Service publication—service descriptions are created in a suitable format and are published according to pre-defined standards in well-known locations;
- Service discovery—information retrieval techniques are employed on the published service descriptions;
- Service selection—results of the discovery process are filtered according to the specified query parameters;
- Service binding—the interface and transport protocol of a service is specified and the service is ready to be executed.

2.6.1 OWL-S: Semantic Markup for Web Services

The Semantic Markup for Web Services (OWL-S) is the OWL ontology for semantic description of web services.[7] OWL-S consists of a service profile for service discovery, a process model which supports composition of services, and a service grounding that associates profile and process concepts with the underlying service interfaces.

Currently, OWL-S Version 1.2 available. The class ServiceProfile of the OWL-S ontology provides a superclass of every type of high-level description of the service. It defines functional properties that describe IOPEs (Input, Output, Preconditions, and Effects) of a service, as well as non-functional properties that describe semi-structured human-readable information for service discovery, e.g. service name, description, and parameters which incorporate further requirements on the service capabilities (e.g., security, quality-of-service, geographical scope, etc.). The class Service model specifies ways of operating the service in a workflow structure with other services.

The service is viewed as a process, which defines the functional properties of the service (IOPEs) together with details of its constituent processes (if the service is a composite service). Functional properties of the service model can be

---

[7]OWL-S 1.2 Release. http://www.ai.sri.com/daml/services/owl-s/1.2/.

shared with the service profile. Interactions between services are represented by service grounding. It enables execution of the Web Service by binding the abstract concepts of the OWL-S profile and process model to concrete message formats and communication protocols. Although different message specifications are supported by OWL-S, the widely accepted Web Services Description Language (WSDL) is preferred as an initial grounding mechanism

2.6.2   The Semantic Annotations for WSDL and XML Schema: SAWSDL

The Semantic Annotations for WSDL and XML Schema (SAWSDL) recommendation[8] defines a set of extension attributes for WSDL, which allows an insertion of semantic descriptions for web services. While the syntactic descriptions of WSDL provide information about the structure of input and output messages of an interface and about how to invoke the service, semantic extension is needed to describe what a web service actually does. The SAWSDL specification defines how semantic annotation is accomplished using references to semantic models, e.g. ontologies. It provides mechanisms by which ontology concepts, typically defined outside the WSDL document, can be referenced from within WSDL and XML Schema components using semantic annotations. The annotation mechanism of SAWSDL uses the abstract definition of services, which is represented in WSDL by Element Declaration, Type Definition, and Interface components. Such a semantic annotation of the abstract part of the service definition consequently enables dynamic discovery, composition, and invocation of services. The extension attributes defined by SAWSDL are as follows:

- the modelReference attribute specifies the association between a WSDL or XML Schema component and a concept in some semantic model;
- the liftingSchemaMapping and loweringSchemaMapping extension attributes are added to XML Schema element declarations and type definitions for specifying mappings between semantic data and XML.

Multiple semantic annotations are allowed for a single WSDL element in service descriptions. Both schema mappings and model references can contain multiple pointers—Uniform Resource Identifiers (URIs) that typically refer to concepts described in an external ontology. Multiple schema mappings are interpreted as alternatives whereas multiple model references are all applied in parallel. SAWSDL does not specify any other relationship between them.

---

[8]Semantic Annotations for WSDL and XML Schema. W3C Recommendation 28 August 2007. http://www.w3.org/TR/sawsdl/.

2.6.3   The Web Service Modelling Ontology: WSMO

The Web Service Modelling Ontology (WSMO) is a conceptual model that was specifically developed to describe semantic web services.[9] The underlying ontological specification of WSMO consists of four major components: ontologies, goals, web services, and mediators. Ontologies provide an agreed common terminology— formal semantics that can be used by all other components. WSMO specifies the following constituents as a part of the description of ontology: concepts, relations, functions, axioms, together with instances of concepts and relations, as well as non-functional properties, imported ontologies, and used mediators.

Goals specify objectives that a client might have when consulting a web service, i.e. functionalities that a web service should provide from the user perspective. The Goal element is characterized by a set of non-functional properties, imported ontologies, used mediators, the requested capability, and the requested WSDL interface. The Web Service elements are described by non-functional properties, references to imported ontologies, used mediators, and the behavioural aspects of web services that are represented by the capability and interface properties. The capability of a web service defines its functionality in terms of preconditions, postconditions, assumptions, and effects, which are expressed by a set of axioms and shared variables. By means of the capability property, a web service may be linked to certain goals that are solved by the web service by means of referenced mediators. The interface of a web service provides further information on how the service functionality is achieved. It describes the behaviour of the service for the client's point of view (i.e., service choreography) as well as the means of achieving overall functionality of the service in terms of cooperation with other services (service orchestration). Mediators represent the elements that enable overcoming structural, semantic, or conceptual mismatches that appear between the components that build up a WSMO description.

All WSMO components are formalized using the Web Service Modelling Language (WSML), which is based on the description logic, first-order logic, and logic programming formalisms.[10] The WSMO framework is supported by the Web Service Modelling eXecution environment (WSMX), which serves as a reference implementation for WSMO.[11]

---

[9]Web Service Modelling Ontology (WSMO). W3C Member Submission 3. June 2005. http://www.w3.org/Submission/WSMO/.

[10]The Web Service Modelling Language WSML: http://www.wsmo.org/wsml/wsml-syntax.

[11]Web Service Modelling eXecution environment. http://www.wsmx.org.

3 Standards Bodies Relevant to the IoT

There are a variety of standards and proprietary platforms for the IoT. At the communication level there are a limited number of standards, including WiFi and ZigBee, and so exchanging data between IoT devices is not a problem. The problem is the discovery and classification of services and the communication at the semantic layer that is summarized under the term machine-tomachine (M2M) communication. Achieving interoperability and establishing services at this level is more challenging and requires semantic knowledge from different domains and the ability to discover and classify services of things in general. This is more difficult to standardize as it changes rapidly and is dependent on particular applications, locations, and use cases. The Alliance for Internet of Things Innovation (AIOTI)[12] has identified a huge range of standards bodies, fora, and consortia that are relevant to IoT.

3.1 British Standards Institute (BSI)

The British Standards Institute (BSI) is relevant to IoT because it developed the PAS 212 HyperCat standard[13] based on the specification from the HyperCat consortium. Use of the standard facilitates the representation and exposure of IoT data hub catalogues over standard web technologies. This improves data discoverability and interoperability, allowing a server to provide a set of resources identified by URIs to a client, each with a set of semantic annotations. As it offers a repository of available resources (nodes, sensors) it might be useful to help semantic devices and service discovery process using ontologies.

3.2 European Telecommunications Standards Institute (ETSI)

ETSI is one of the three European Standards Organizations (ESOs) formally recognized by the European Commission (EC) as providing European Standards (ENs). It develops standards predominantly in the telecommunications area but has recently moved "up the stack" and is now developing standards and architectures for M2M and ITS. The following technical committees (TCs) and Industry Specification Groups (ISGs) are relevant to the IoT: ETSI TC SmartM2M was set up to develop specifications for M2M services and applications focussing on IoT and Smart Cities. It primarily supports European policy and regulatory requirements including

---

[12]https://aioti.eu/.

[13]BSI PAS 212 "Automatic resource discovery for the Internet of Things-Specification"—http://shop.bsigroup.com/ProductDetail/?pid=000000000030327418.

mandates in the area of M2M and IoT. It identifies European Union (EU) policy and regulatory requirements on M2M services and applications to be developed by oneM2M, and the conversion of the oneM2M specifications into European Standards. TC SmartM2M has mapped SAREF to the oneM2M Base Ontology and is also evolving SAREF and extending it to different domains. ISG CIM has developed a Context Information Management layer which sits above the oneM2M Service Layer and is used to integrate data for IoT applications. It has developed standards on APIs, Data Publication Platforms, and the Information Model.

3.3   oneM2M Partnership Project

The oneM2M Partnership Project involves 7 regional SDOs as "Type 1" partners:

- Association of Radio Industries and Businesses (ARIB),
- Alliance for Telecommunications Industry Solutions (ATIS),
- China Communications Standards Association (CCSA),
- European Telecommunications Standards Institute (ETSI),
- Telecommunications Industry Association (TIA),
- Telecommunications Standards Development Society, India (TSDSI),
- Telecommunications Technology Association (TTA),
- Telecommunication Technology Committee (TTC).

This project, therefore, has an international scope and any member of one of these regional SDOs can participate fully in oneM2M. The structure of oneM2M can be found here.[14] The work on semantics/ontologies is carried out in WG MAS (Management Abstraction and Semantics). This is developing a Base Ontology based on the requirements of specific ontologies such as SAREF. Stages 1 and 2 have already been published as an ETSI TS. Stage 3 was finalized in August 2016 as part of oneM2M.[15]

3.4   Institute of Electrical and Electronics Engineers (IEEE)

The Standards Association of the Institute of Electrical and Electronics Engineers (IEEE-SA) has established a reference framework and architecture for IoT defined in IEEE 2413. This aims to promote cross-domain interaction, aid system interoperability and functional compatibility across IoT systems. IEEE-SA also develops IoT standards across different domains:

---

[14] www.onem2m.org/about-onem2m/organisation-and-structure.

[15] http://www.onem2m.org/technical/published-drafts.

- Communications (IEEE 802—wireless/wireline standards, IEEE 1901 on BPL),
- Transportation (IEEE 802.11p, IEEE 1609P),
- Smart Grid standards and Smart Energy Profile (IEEE 2030.5),
- Sensor Standards (IEEE 1451, IEEE 2700).

Standards relevant to the IoT include IEEE 802 LAN/MAN Standards[16] and the Suggested Upper Merged Ontology (SUMO).

3.5 Internet Engineering Task Force (IETF)

The Internet Engineering Task Force (IETF) has developed the following standards relevant to the IoT:

- Constrained Application Protocol (CoAP) which is a protocol for device communication over the Internet.
- 6LoWPAN for constrained radio links.
- ROLL which is a routing protocol for constrained-node networks.

Figure 6.1 shows a layered view of the IETF IoT protocol stack. Applications and devices are interconnected via IPv6 which has sufficient address space to be available on a wide variety of different devices and link layer technologies, each of which are tailored to meet the specifics demands of their domain: wired vs wireless, short vs long range, line of sight vs non LoS, high vs low data throughput, narrowband vs wide band, etc.

The CoAP protocol specification was developed by the Constrained RESTful Environments (CoRE) WG and published as RFC 7252 in June 2014. CoAP uses the same RESTful principles as HTTP, but is much lighter so that it can be run on constrained devices. To achieve this, CoAP has a much lower header overhead and parsing complexity than HTTP. It uses a 4-bytes base binary header that may be followed by compact binary options and payload.

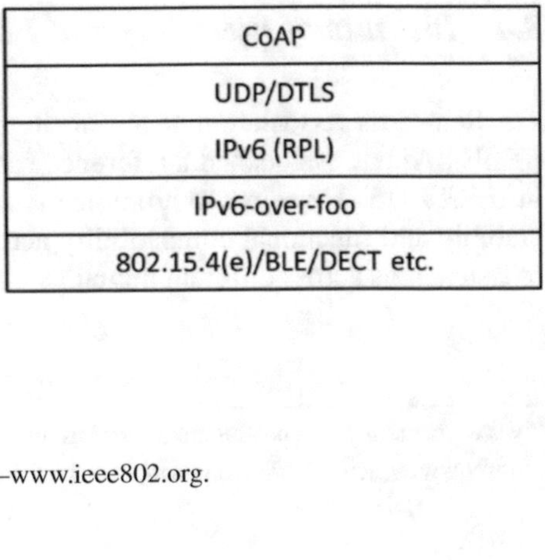

Fig. 6.1 IETF IoT protocol stacks

CoAP
UDP/DTLS
IPv6 (RPL)
IPv6-over-foo
802.15.4(e)/BLE/DECT etc.

---

[16]IEEE 802 LAN/MAN Standards Committee—www.ieee802.org.

3.6   ISO/IEC JTC1 Information Technology

The International Organization for Standardization (ISO) is an international standards organization with worldwide representation. The full list of ISO technical committees is available online[17] but the main activity of relevance to the IoT is JTC1, a joint technical committee of ISO and the International Electrotechnical Commission (IEC) and provides the standards approval environment for integrating diverse and complex ICT technologies. It has a new WG10 devoted to the IoT and also develops standards for Identity Management and Privacy Technologies. JTC1/WG10 developed an IoT Reference Architecture (ISO/IEC 30141) that defines reference models and architectural views which can be easily extended to a real architecture. Also, Technical Report on IoT Use Cases will be continuously updated to collect various use cases including interoperability, smart manufacturing, and smart wearable devices. Support for interoperable IoT systems is becoming more important and WG 10 will develop standards for interoperability for IoT Systems (ISO/IEC 21823-1: Framework, ISO/IEC NP 21823-x(2): Semantic interoperability, and ISO/IEC NP 21823-x(3): Network connectivity). WG 10 now considers wearable technologies as one of its key workstreams. For Systems Integration, WG 10 will keep cooperation with JTC 1 entities whose working items are related with IoT and wearable technologies as well as outside JTC 1. It is also working on IoT use cases based on WG10_N0090 template including:

- IoT Use Cases in area of Ambient Assistive Living (AAL),
- A Smart Glasses use case from MPEG,
- Smart Wearable Device use cases on "Searching for people with cognitive impairment" and "Sleep Monitoring System",
- Intelligent transport systems, Smart Parking and eHealth.

JTC1/WG7 has issued standards for a Sensor Network Reference Architecture (SNRA) (ISO/IEC 29182) multipart standard; Part1: General overview and requirements; Part 2: Vocabulary and Terminology; Part 3: Reference Architecture Views; Part 4: Entity models; Part 5: Interface definitions; Part 6: Applications; Part 7: Interoperability guidelines. ISO/IEC 20922:2016 is a Client-Server publish/subscribe messaging transport protocol. It is lightweight, open, simple and designed so as to be easy to implement. These characteristics make it ideal for use in many situations, including constrained environments such as for communication in M2M and IoT contexts where a small code footprint is required and/or network bandwidth is at a premium ISO/IEC 14543 covers Home and Building Electronic Systems (HBES).

ISO/IEC SC27 covers Development of standards for the protection of information and ICT. This includes generic methods, techniques, and guidelines to address both security and privacy aspects, such as:

- Security requirements capture methodology;

---

[17] http://www.iso.org/iso/home/standards_development/list_of_iso_technical_committees.htm.

- Management of information and ICT security; in particular information security management system (ISMS) standards, security processes, security controls and services;
- Cryptographic and other security mechanisms, including but not limited to mechanisms for protecting the accountability, availability, integrity, and confidentiality of information;
- Security management supports documentation including terminology, guidelines as well as procedures for the registration of security components;
- Security aspects of identity management, biometrics and privacy;
- Conformance assessment, accreditation and auditing requirements in the area of information security management systems;
- Security evaluation criteria and methodology.

SC27 engages in active liaison and collaboration with appropriate bodies to ensure proper development and application of SC 27 standards and technical reports in relevant areas. The following working groups are relevant to the IoT architecture:

- WG 1—Information Security Management Systems.
- WG 2—Cryptography and Security Mechanisms
- WG 3—Security Evaluation, Testing and Specification
- WG 4—Security Controls and Services
- WG 5—Identity Management and Privacy Technologies

MQTT[18] is a Client-Server publish/subscribe messaging transport protocol, now published by OASIS.[19] It is lightweight, open, simple and designed to be easy to implement. These characteristics make it ideal for use in many situations, including constrained environments such as for communication in IoT contexts where a small code footprint is required and/or network bandwidth is at a premium.

3.7 *International Telecommunications Union (ITU)*

The standardization sector of the ITU (ITU-T) has over 700 member organizations including the ministries of communication from most countries. ITU SG20 "IoT and its applications including smart cities and communities" develops standards for IoT interoperability including APIs. SG20 has been defining key performance indicators for the performance of smart cities, which include metrics for the efficiency of services such as smart grids, e-health, e-transport, and open data. A new Work Item on a semantic ontology model for IoT is being proposed for worldwide acceptance based upon ETSI TR 101 584. A new Work Item on "Requirements and capabilities for common IoT service discovery through IoT gateway in the IoT environments"

---

[18]MQTT: http://mqtt.org/.

[19]https://www.oasis-open.org/news/announcements/mqtt-version-3-1-1-becomes-an-oasis-standard.

has been proposed. A new Work Item is proposed to study IPv6 potential and impact on the Internet of Things and smart cities and communities.

A Focus Group on Data Processing and Management (FG-DPM) was created in 2017 to address data processing for the IoT. It completed its work in 2019 and developed standards on:

- Data processing and management framework for IoT and smart Cities and communities
- Web based data model for IoT and smart city
- SensorThings API—Sensing
- Framework to support data interoperability in IoT environments
- Overview of blockchain for supporting IoT and SC&C in DPM aspects
- Blockchain-based data exchange and sharing for supporting IoT and SC&C
- Blockchain-based data management for supporting IoT and SC&C
- Identity framework in blockchain to support DPM for IoT and SC&C
- Framework for security, privacy, risk, and governance in data processing and management
- Overview of technical enablers for trusted data
- Framework to support data quality management in IoT
- Data economy: commercialization, ecosystem and impact assessment

3.8   Open Connectivity Foundation (OCF)

The OCF now encompasses the IoTivity and AllJoyn projects and the activities of the previous AllSeen Alliance[20] and Open Interconnect Consortium (OIC).[21] The AllSeen Alliance developed AllJoyn, an open source software framework that makes it easy for devices and apps to discover and communicate with each other. Developers can write applications for interoperability regardless of transport layer, manufacturer, and without the need for Internet access. The software has been and will continue to be openly available for developers to download and runs on popular platforms such as Linux and Linux-based Android, iOS, and Windows, including many other lightweight real-time operating systems. The first version of IoTivity specification was published in September 2015 by the OIC. It is based around CoAP, the IETF protocol for device communication over Internet. They do not use ontologies but their data models are defined using RAML (a language for describing RESTful APIs).

---

[20] AllSeen Alliance—www.allseenalliance.org.

[21] Open Interconnect Consortium: https://openconnectivity.org.

3.8.1 The Thread Group

The Thread Group[22] was designed to create an effective way to connect and control products in the home. The following key features are available:

- Simple for consumers to use
- Always secure
- Power-efficient
- An open protocol that carries IPv6 natively
- Based on a robust mesh network with no single point of failure
- Runs over standard 802.15.4 radios
- Designed to support a wide variety of products for the home: appliances, access control, climate control, energy management, lighting, safety, and security

3.9 World Wide Web Consortium (W3C)

The World Wide Web Consortium (W3C) is the main standards organization developing standards for the Web. The most relevant to the IoT are:

- W3C/OGC Spatial Data on the Web (SDW) WG. The goal of the SDW WG is to clarify and formalize standards for the representation of spatio-temporal data, including data coming from sensors. This WG is explicitly chartered to work in collaboration with the Open Geospatial Consortium (OGC), in particular, the Spatial Data on the Web Task Force of the Geosemantics Domain WG. Among other deliverables, the SDW WG will standardize updated versions of the Time and the Semantic Sensor Network (SSN) ontologies, previously defined in the scope of the W3C, and will provide best practices for publishing and using spatial data on the Web.
- W3C Web of Things (WoT) IG. Their goal is to identify requirements for the technology building blocks for the application layer that forms the WoT, with the idea of reaching out and collaborating with interested parties to create a new W3C WG.
- W3C Linked Building Data (LBD) Community Group. The LBD CG aims to define existing and future use cases and requirements for linked data based applications across the life cycle of buildings. The group has a focus on providing ontologies for representing Building Information Models following the IFC data model (an official ISO standard—ISO 16739:2013).

---

[22]The Thread Group—www.threadgroup.org.

3.10 ZigBee Alliance

The ZigBee Alliance provides a foundation for the IoT and is well established in the AAL communications protocols area. In conjunction with the HomePlug Powerline Alliance it has developed Smart Energy Profile (SEP) 2.0 (e.g., for light bulbs). ZigBee is not just about Transport and provides an example of creating Interoperability as a service.

4 Standards for IoT Applications

There are many thousands of IoT applications operating around the world at a range of maturity levels and in a wide range of domains. The most common are in the domains of eHealth and Assisted Living and in Smart Transport and Personal Mobility. Standards for the following domains are covered in this section:

- Smart homes and buildings
- Smart energy
- Smart transport and personal mobility
- eHealth and assisted living

In addition, standards related to Security and Privacy are considered as these are required across all domains.

4.1 Smart Homes and Buildings

From an information perspective, the building industry is highly complex. A building is usually a one-off design and is difficult to produce using industrial approaches. Builders and property developers focus primarily on building as cheaply as possible, and usually give little attention to the operational phase and maintenance of the building. An important objective of the IoT is to enable a smarter homes in terms of a more efficient and more open use of the information and services provided by smart appliances. By gathering data from different household (smart) appliances, in particular from sensors that are incorporated in them, energy efficiency will be improved. A platform such as VICINITY will go beyond this to create ways to open the data and services of the smart appliances to independent service operators. The availability of the data from smart appliances will pave the way to new services in other domains such as security, e-health, and transport that will have an impact on individuals and the community as a whole. The standards that Smart Appliances rely on include:

- ETSI Smart Appliances REFerence ontology (SAREF)
- ETSI TS 103 267 Smart Appliances; Communication Framework.

- ETSI TS 103 264 Smart Appliances Common Ontology and oneM2M Mapping.

CEN TC 247 covers Technical Building Management, Automation and Control and many of the standards are relevant to Smart Energy applications. CENELEC TC 205 covers Home and Building Electronic Systems (HBES). EN 50090 is the series of European standards for home and building control. In addition, the following CENELEC standards are relevant to smart appliances:

- CENELEC EN 50523-1:2009: Household Appliances Interworking—Part 1: Functional Specification.
- CENELEC EN50523-2:2009: Household Appliances Interworking—Part 2: Data structures.

ISO TC 184 sub-committee 4 (SC4) Modelling of industrial, technical, and scientific data to support electronic communication and commerce is dealing with standards for buildings. ISO 16739:2013 specifies a conceptual data schema and an exchange file format for Building Information Model (BIM) data. The conceptual schema is defined in EXPRESS data specification language. The standard exchange file format for exchanging and sharing data according to the conceptual schema is using the *Clear text encoding* of the exchange structure. Alternative exchange file formats can be used if they conform to the conceptual schema.

ISO 16739:2013 represents an open international standard for BIM data that is exchanged and shared among software applications used by the various participants in a building construction or facility management project.

ISO 16739:2013 consists of the data schema, represented as an EXPRESS schema specification, and reference data, represented as definitions of property and quantity names and descriptions.

A subset of the data schema and referenced data is referred to as a model view definition. A particular model view definition is defined to support one or many recognized workflows in the building construction and facility management industry sector. Each workflow identifies data exchange requirements for software applications. Conforming software applications need to identity the model view definition they conform to.

The following are within the scope of ISO 16739:2013: BIM exchange format definitions that are required during the life cycle phases of buildings: demonstrating the need; conception of need; outline feasibility; substantive feasibility study and outline financial authority; outline conceptual design; full conceptual design; coordinated design; procurement and full financial authority; production information; construction; operation and maintenance.

4.2 Smart Energy

Smart Energy includes the areas of smart grids and electric vehicle charging.

IEC 62196 Conductive charging of electric vehicles defines a set of electrical connectors for electric vehicles. This is based on IEC 61851 Electric vehicle conductive charging system which establishes general characteristics, including charging modes and connection configurations, and requirements for specific implementations (including safety requirements) of both EV and EVSE in a charging system. For example, it specifies mechanisms such that, first, power is not supplied unless a vehicle is connected and, second, the vehicle is immobilized while still connected. The Smart Energy Demand Coalition (SEDC)[23] is a European industry association dedicated to making the demand side a smart and interactive part of the energy value chain. Their focus is to promote demand-centred programs within the areas of Demand Response, energy usage feedback and information, smart home, in-home and in-building automation that form the heart of the Smart Grid. Their vision is a strong participation of the demand side in the European electricity markets, which will lead to the long-term goals of the Smart Grid: affordability, security of supply, and reduced carbon emissions. The purpose of the Coalition, a not-for-profit organization, is to create a community of expertise on demand side programs and their role in creating efficient electricity markets. Through this community, the Coalition will bring forward useful information on price responsive loads, program and technology experience, market structures, and market rules. This will include information on market participants' roles, consumer needs and actions, enabling technologies, and specific programmes as appropriate, focussing efforts in five areas:

- Interact with policymakers on behalf of the Demand Side Program industry
- Intelligence and information gathering for SEDC Members
- Giving members of the SEDC greater visibility
- Working with trade, financial, and general media to raise awareness of Demand Response, feedback, smart home, etc.
- Providing opportunities for networking and partnering among demand side programme and smart grid companies

The Universal Smart Energy Framework (USEF) is a standards initiative on smart interconnected energy. This has developed the foundations of one integrated system which benefits all players—new and traditional energy companies and consumers.

4.3 Smart Transport and Personal Mobility

Smart Transport includes the areas of Smart Parking, Intelligent routing and Congestion control, Smart Logistics, and Personal Mobility. CEN TC 278 is heavily integrated with ISO TC 204 and covers Transport Telematics and Traffic, which may be relevant to Smart Grid and Parking applications. IoT in Transport is still not

---

[23] Smart Energy Demand Coalition (SEDC): http://www.smartenergydemand.eu/.

covered as they do not see connected vehicles as part of the IoT and so do not see the need for open communications except to support specific applications.

ISO/TS 21219 is a multipart standards covering traffic and travel information and Part 14 is under development which will cover Parking Information (TPEG2-PKI), Weather Information, Geographic Referencing, Traffic Flow and Prediction Appliances. ETSI Intelligent Transport Systems (ITS) is also relevant but does not yet see connected vehicles as "devices" in the IoT and hence, do not see the need for open communications except to support specific applications.

4.4   eHealth and Assisted Living

The eHealth domain primarily covers Smart eHealth and Assistive Living at home, also Fitness Tracking and Preventive Medicine. Smart homes contains health sensors linked to IoT gateways which provide information to web servers. Standards on which data and information exchange will rely include:

* Integrating the Healthcare Enterprise (IHE) Profiles[24];
* Health Level 7 International (HL7)[25] Standards;
* The DICOM Standard[26];
* Continua Health Alliance.[27]

IHE[28] is an initiative by healthcare professionals and industry to improve the way computer systems in healthcare share information. IHE promotes the coordinated use of established standards such as DICOM and HL7 to address specific clinical needs in support of optimal patient care. Systems developed in accordance with IHE communicate with one another better, are easier to implement, and enable care providers to use information more effectively.

CEN TC 204 covers Medical Devices. Many of the standards are relevant to IoT eHealth applications. CEN TC 251 covers Health Informatics which is also relevant to IoT eHealth applications. The following standards are particularly relevant:

* ISO/FDIS 25237:2016 Health informatics—Pseudonymization
* ISO/IEEE 11073: Health Informatics: Personal health device communication: Device specialization.
* CEN-TC251_N2016076 New Work Item proposal on Health and Wellness apps.
* prEN/ISO/FV 27799—Health informatics—Information security management in health using ISO/IEC 27002 (ISO/FDIS 27799:2016).

---

[24]IHE Profiles: http://www.ihe.net/Profiles.

[25]Health Level 7 Standards: https://www.hl7.org.

[26]The DICOM Standard: http://dicom.nema.org/standard.html.

[27]Continua Health Alliance: http://www.continuaalliance.org.

[28]Integrating the Healthcare Enterprise: http://www.ihe.net/.

The Continua Health Alliance[29] addresses Personal Connected Health and is required for IoT eHealth applications. Continua includes over 100 industry leading companies and healthcare organizations worldwide. Implementations of Continua specifications in the upcoming area of welfare technologies are popular with European Governments and interest from vendors is growing.

The Continua Health Alliance standard addresses the fundamentals of data exchange between medical devices. Use of Continua-enabled products in IoT eHealth applications will provide end-users with increased assurance of inter-operability between devices and enabling them to easily share information with caregivers and service providers. Continua Design Guidelines provide a flexible implementation framework for authentic interoperability, containing references to the standards and specifications that Continua selected for ensuring interoperability of devices. Some of the standards selected are:

- Bluetooth for wireless and USB for wired device connection,
- ISO/IEEE 11073 Personal Health Data (PHD) Standards
- The IHE (originally IETF) External Data Representation (XDR) standard for the exchange of clinical documents between healthcare enterprises

Continua could provide a standardized way of obtaining information from medical devices although more work would be needed to determine the protocols that would be used. IEC 62304 Medical device software—software life cycle processes specify life cycle requirements for the development of medical software and software within medical devices. It is harmonized by the EU and the USA and therefore can be used as a benchmark to comply with regulatory requirements from both these markets. ISO TC 215 covers Health Informatics. Relevant work is the ISO/NP TS 11633-1 Health Informatics, Info security management for remote maintenance of medical devices and MIS—Part 1 Requirements and risk analysis.

4.5   Security

IoT will not create an improvement in peoples' lives unless their concerns over accessibility, data protection, security, and privacy are addressed. Standards are critical to these issues as well as for interoperability. Topics relevant to Security include Identity Management, Anonymity and Pseudonymity, Credentials and Attributes and Access Management. Adequate security for the IoT should consider the following standards:

- ISO Security Framework;
- ITU-T Security in Telecommunications and Information Technology: An overview of issues and the deployment of existing ITU-T Recommendations for secure telecommunications;

---

[29]Continua Alliance http://www.continuaalliance.org/about-the-alliance/join.

- EN 61508 Functional safety;
- ISO 9160 Data encipherment—Physical layer interoperability requirements;
- IEEE 802.11 Security of wireless communication networks;
- IETF RFC 2818 HTTP Over TLS[30];

End-to-end security should be included in the communications layer in order to provide adequate security for information in the IoT. However, a key issue is the difficulty of achieving sub-mS end-to-end-response times for exchange of data if encryption and decryption are to be used by IoT devices, especially if using a Trusted Third Party (TTP) architecture. This is not possible using current technologies. Organizations working on IoT security include:

- Cloud Security Alliance (CSA)
- ETSI TC Cyber
- ISO/IEC JTC 1/WG 10 on IoT
- ITU-T SG 20 ETSI TC Cyber is addressing topics such as:
- Cyber Security
- Security of infrastructures, devices, services, and protocols
- Security advice, guidance and operational security requirements to users, manu-facturers, and network and infrastructure operators
- Security tools and techniques to ensure security
- Creation of security specifications and alignment with work done in other TCs

The importance of Privacy to the IoT has been raised by the recent EC General Data Protection Regulation (GDPR)[31] and the main impacts are described here.[32] Topics relevant to Privacy include a privacy framework, a privacy reference architecture, privacy infrastructures, privacy impact assessment, and specific privacy enhancing technologies (PETs). ISO/IEC JTC1/SC27/WG5 "Identity Management and Privacy Technologies" covers the development and maintenance of standards and guidelines addressing security aspects of identity management, biometrics, and the protection of personal data.

5   The Future of IoT Standards

In the future, IoT standardization must answer the following questions:

- Will a standardized approach be sufficient to ensure interoperability between IoT platforms at all levels?
- Can semantic ontologies ever be sufficiently detailed to ensure interoperability at higher levels?

---

[30]IETF RFC 2818 HTTP Over TLS http://tools.ietf.org/html/rfc2818.

[31]EU General Data Protection Regulation (GDPR) home page: http://www.eugdpr.org/.

[32]https://iapp.org/resources/article/top-10-operational-impacts-of-the-gdpr/.

- How can context information be used to ensure that the "right" meaning is attached to a given object or quality to allow it to be used effectively?
- How can standards help to provide effective action against threats to security and privacy in the IoT?

IoT is still a growing area and many fora and consortia have been created to look at the various requirements and gaps in standards for the IoT. These are very specific to their own area of expertise and in many cases are developing their own specifications and market requirements. For these reasons, it becomes more and more important to support interoperability between different levels of IoT systems in various IoT platforms based on different standards. The lower levels are already well covered by different protocols and ways to interoperate between them have been established (e.g., via gateways). However, there are still problems of interoperation at the higher layers, especially the semantic layer still has many difficulties to establish "meaning" between different IoT implementations. SAREF has been extended to include energy, buildings, environment, smart cities, industry, and agriculture and is being extended further to cover automotive, health and ageing well, wearables, and water.

In future, the deployment of many smart city use cases designed to improve the well-being of citizens and allow cities to be managed effectively, will depend on the availability of horizontal IoT platforms. These will enable cross-domain applications to be deployed that will address city priorities such as improvements in air quality, traffic congestion, energy availability, health and assisted living. However, all of these are critically dependent on the further development of IoT standards that will enable the required interoperability between platforms, devices, gateways, servers, and applications in a multi-provider, multi-stakeholder world.

Chapter 7
Security and Trust

Viktor Oravec

1 Introduction

Security and trust are topics which are extremely important for the Internet of Things. While the smart devices became part of our lives, secure usage and trust in the IoT is crucial. The IoT systems usually follow the fog or edge computing architecture with distributed or even decentralised concept. Due to this mixture of architectures, requirements and measures, required to settle good security principles, are more complex than in traditional web application development. However, good news is that most of the security procedures, approaches, concepts and tools can be reused from well defined ICT best practices [4].

The process of making the secure IoT ecosystem starts with building the security framework that will be discussed in Sect. 2. Then, the next step is security stakeholders analysis and identification of the stakeholders' security goals followed by the selection of assets that need to be secured. These steps can give us solid understanding what parts of a system must pass through the security process. After that, possible vulnerabilities and threats need to be identified and security risk log needs to be built.

When the security risk log is built, it is important to identify the organisation or technical security measures to mitigate these risks to the acceptable level. It is important to note that the built security framework only mitigates identified risks, but does not remove them. The question to be asked here is *What is the acceptable level for a risk?* The acceptable level is defined by the stakeholder raising the particular security issue. The rest of the chapter gives the overview on how to tackle the security and trust in IoT domain and to understand the basic concept for building

V. Oravec (✉)
bAvenir s.r.o., Bratislava, Slovakia
e-mail: viktor.oravec@bavenir.eu

© Springer Nature Switzerland AG 2021
C. Zivkovic et al. (eds.), *IoT Platforms, Use Cases, Privacy, and Business Models*,
https://doi.org/10.1007/978-3-030-45316-9_7

a full security framework which will be enhanced with selected organisational and technological measures based on the IoT solutions properties.

2   Security Framework

The goal of the security framework is to establish the security requirements baseline to mitigate the security risk. The framework views the security of the IoT Platform from the perspective of its stakeholders with particular security goals throughout the whole product value chain. This chapter provides a guide through the whole process of building a security framework, with illustrative examples from the VICINITY Platform Security Framework. Note, the VICINITY Platform Security framework is introduced for demonstration purpose, in order to obtain a better view on how a security framework can be built.

2.1   IoT Ecosystem Stakeholders and Their Role in Security Model

The security needs are evaluated from the IoT ecosystem's stakeholders point of view. The list of stakeholders includes any entity who is involved in the whole product life-cycle and value chain. The most common mistake is to involve only providers of the IoT ecosystem and its users. However, it is necessary to go beyond that. For example, if we take into account the Internet of things reference architecture [2], at least nine functional domains need to emerge, which needs to be considered during the stakeholders analysis. Good news is that the security framework needs to be tailored to the needs of particular IoT project and usually not all parts of the reference architectures are relevant for the solution. A good example of tailor view on stakeholders is the Web of Things stakeholders analysis where only part of the IoT reference architecture is relevant and hence covered [14].

It is common to overlook parts such as IoT ecosystem management functional domain which deals with the configuration management, monitoring and other supporting functionalities of the IoT ecosystem provision. Here, roles like configuration engineers or DevOps engineers can have significant impact from business continuity to the system infrastructure vulnerability. Another part which system engineers take as relevant is communication layer. The communication protocols such as Wi-Fi, Ethernet, XG and LoRaWAN are usually reused and integrated into projects; however, a closer look on who is providing the layer and how the layer is provided needs to be taken into account.

2.1.1 Stakeholders Analysis

Here we will show how critical stakeholders can be identified using the VICINITY example. The first step is identification of functional roles in an ecosystem. In the VICINITY platform several functional roles were identified:

- Internet of Things operator
- Device owner
- Service provider
- System integrator

From the VICINITY operation point of view the following most critical stakeholders can be identified:

- bAvenir, s.r.o.—as VICINITY Platform operator.
- Microsoft—Provides and supports NodeJS framework.
- Google—Provides and supports AngularJS framework.
- MongoDB Inc.—Provides and supports MongoDB framework.
- Hetzner GmbH—Provides the cloud infrastructure for the VICINITY Platform.
- Docker Inc.—Provides the container features for the VICINITY Web Applications.

Some of these stakeholders have direct impact on security such as providers of the cloud infrastructures and application running environments. On the other hand, providers of implementation frameworks do not have direct impact; however, the framework needs to be monitored for the security vulnerabilities during implementation of operational measures.

2.2 IoT Ecosystem Asset Identification

The next step in building the security framework is to identify the assets which need to be protected. Again the IoT reference architecture can help in the identification of these assets. The common assets are devices, IoT gateways, services, collected data, aggregated data, service products, etc. For each identified asset the *Failure Mode and Effect Analysis* needs to be performed. It will help us identify the failure of the assets for final product or process, assess the risk associated with those failure modes, rank the issue's importance and carry out corrective actions. In general, if we want to protect something valuable (in this case asset), we need to determine how risky certain situations are, how probable these are and what is the balance between countermeasures and potential impact of threat if it happens.

2.2.1   IoT Ecosystem Assets Life-Cycle and Their Impact on Security

It is not enough to check assets existence, but it is important to analyse its whole life-cycle and evaluate its impact from security point of view. For example, device life-cycle is a complex process including stages like manufacturing, assembling, selling, distribution, usage, maintenance, decommissioning, etc. [14, 15]. Each stage includes one or more stakeholders. To reduce the complexity of security framework it is important to identify which parts of the asset life-cycle should be covered. In particular cases, like VICINITY Platform, the manufacturing, selling or distribution stages are not relevant for the platform. However, it is important to cover parts of the device operational life-cycle from point where the device is being registered in the VICINITY (deployment of supported device or integration of integrated IoT platform) until its removal from the platform.

VICINITY Showcase   Based on the previously mentioned VICINITY Stakeholder analysis the following assets need to be protected from the VICINITY Neighbourhood Manager's point of view:

- *Primary assets*: User profiles and services

    - Data exchange between parties
    - Registration of devices and services
    - Sharing of devices and services between parties
    - Device profile (Thing descriptions)
    - Service profile (Thing descriptions)
    - Device and service access rules
    - User and organisation profiles
    - IoT infrastructure access points

- *Secondary assets*: Components of an IoT platform. Concretely, for the example of the VICINITY platform these components are as follows:

    - VICINITY Neighbourhood Manager
    - VICINITY Communication Server
    - VICINITY Peer-to-peer network
    - VICINITY Gateway API

Each of these assets need to pass through the *Failure Mode and Effect Analysis* for the evaluation of confidentiality, integrity and availability. Table 7.1 demonstrates one example based on one primary and one secondary asset identified in the VICINITY project.

Note   For more details on the VICINITY components we refer a reader to Chap. 9 that provides a step-by-step instruction guide how to master the IoT with the VICINITY platform.

Table 7.1 Example of the selected primary and secondary asset analysis

Asset	Failure mode	Effect analysis
Data exchange between parties	Data could not be exchanged between parties, data can be disclosed to 3rd party (eavesdropped)	Value-added services business continuity and/or (business) privacy breach
VICINITY Neighbourhood Manager	The device and service registry not accessible, accessed by unauthorised personnel, the registry is corrupted	Sharing rules could not be changed, sharing rules and registry for particular organisation is compromised

2.3 Asset Vulnerabilities and Threat Identification

For each asset a set of potential vulnerabilities and threats needs to be identified. When using the term 'threat', we shall clearly define what threat and threat probability mean. Generally, the term 'threat' can be described as: *Anything that can exploit a vulnerability, intentionally or accidentally, and obtain, damage, or destroy an asset.* Several vulnerability and threat categories are included in ISO/IEC 27005:2011 [10] or in the extended list for the OWASP Internet of Things Vulnerabilities [5]. For example, for the asset *Data exchange between parties* the set of vulnerabilities can be identified with associated threats, listed in Table 7.2. Assets, especially device and software components, can change their properties during their life-cycle, such as the functional updates and security assessments. Thus, it is important to repeat this step in the security framework regularly, as a part of the security processes in the organisation.

2.4 Risk Analysis and Management

The goal of the risk analysis is to put together all security analysis results, namely identified roles and their security requirements, the list of assets and their vulnerabilities and threats. The list of the identified risks needs to be created and their impact needs to be assessed. The security risk log is similar to a project risk log; the most important risks are those with the highest impact and the highest probability (cf. Table 7.3). The risks above acceptance thresholds need to be addressed by the organisational or technical measures. The risks below the acceptance thresholds are kept in the risk log for the latter assessment. It is important to be aware that risk impact and probability can change over time. Hence, regular risk analysis needs to be performed, in order to identify these movements, especially ones whose impact and probability levels cross the acceptance thresholds. It may be important to identify situations when the risk moves from values above the threshold to values below it. When security measures are not relevant anymore either based on security risk analysis or based on the nature of the measure, they can be removed.

Table 7.2 VICINITY example of the vulnerabilities, threats and surface of attack analysis

Vulnerability	Threat	Surface of attack
Wrong allocation of rights in the VICINITY Neighbourhood Manager	Abuse of rights	VICINITY Neighbourhood Manager, VICINITY User's organisation
Poor password management	Forging rights	VICINITY Neighbourhood manager, VICINITY Adapter and VICINITY User's organisation
Unprotected password tables	Forging rights	VICINITY Neighbourhood manager, VICINITY Adapter
Uncontrolled use of software	Tampering with software	VICINITY Gateway API Docker Image, Java Library and source code
Unprotected communication lines	Eavesdropping	Communication between VICINITY Adapter—Gateway API, Communication in VICINITY Peer-to-peer network, VICINITY Gateway API and Communication Server
Unprotected sensitive traffic	Eavesdropping	Communication between VICINITY Adapter—Gateway API, Communication in VICINITY Peer-to-peer network, VICINITY Gateway API and Communication Server

This is important in cases when this measure creates the unnecessary waste in a product/service life-cycle. The typical example for this situation is when legislation or standardisation rules are changed, which results in simplification of security procedures (Fig. 7.1).

Attack Surfaces in IoT Ecosystem and Identification of the Security Boundaries
After the whole risk analysis the next step is to provide technical solutions for applying certain security measures.

Each identified threat is directed towards a technical solution and this point is called attack surface. Different attack surface taxonomies are useful in different IoT context such as IoT ecosystems [6], smart home installations [7], road vehicles [9] and cloud services [8].

Table 7.3 VICINITY example of the risk analysis

Risk	Compromise of exchange data between parties by tempering with the VICINITY Gateway API and VICINITY Adapter
Threats impact	
Level	*Assessment*
Medium	Assets might be disabled, not damaged. Most serious impact is in compromise of trust in validity and reliability of primary asset (services). Compromised or untrusted service threatens basics of projects
Threats probability	
Level	*Assessment*
High	Due to spread infrastructure with a lot of interconnected devices the probability of compromise is high
Risk assessment	This risk shall be considered as probable and because of its medium impact evaluation and high probability. Risk must be protected against with relevant countermeasures taken, periodically checked and audited
Security measure to reduce the risk	The VICINITY Gateway API JARs and Docker images need to be protected by finger printing. SDLC of VICINITY API components needs to be secured on organisational level

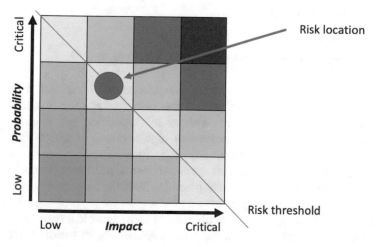

Fig. 7.1 Risk assessment matrix

3 Security and Privacy Measures

Based on the risk analysis relevant security and privacy measures have to be applied for particular failure modes in order to reduce the impact of exploitation of particular vulnerabilities and threats to an acceptable level. The measures can be divided into the following groups:

- Organisational measures

- Technical measures
- Data privacy measures

3.1 Organisational Measures

The organisational measures improve organisation's processes, data flow principles and human resources responsibilities. These measures need to be defined in an organisation's security policy. Security policy can be divided into several levels:

- *Organisational level*—the security policy that covers the common measures for a whole organisation. The security policy considers strategic and business department requirements.
- *Project level*—the security policy that is complementary to organisational one, however, addresses project's customer requirements.
- *Product level*—the security policy that is complementary to organisational one, however, addresses specifications for particular products.

In small organisations such as SMEs and start-ups these levels are usually merged into one simple security policy.

3.1.1 Impacts of Security Measures on the DevOps

The organisational measure can make the processes slower and more work-intensive. Thus, it is important to find the balance between business (where the money is earned) and security (where the reputation can be harmed). In commoditisation and cloud era, process of high amounts of data with high and global availability can be a difficult task. It is even difficult, when delivery time of software and/or hardware features spans between 2 and 12 weeks cycles. These delivery processes have to be as lean as possible. Thus, the security measures need to go hand in hand with DevOps principles and processes for maximising flow through value stream, while keeping the security risks at an acceptable level.

From the DevOps point of view the most sensitive security measures are:

- *Supplier security requirements policy*—suppliers of hardware and software in IoT initiatives should fulfil different security requirements. However, these requirements can slowdown the assets provision (i.e., provision of new version devices) which can imbalance the value stream and creates unnecessary waiting queues.
- *Business continuity*—keeping a healthy value stream supports the business continuity for services and products. Feedback from the technical infrastructure enables to spot or even predicts the problems which can be solved beforehand.

- *Cyber security incidents management*—should be inline with DevOps process. In that case cyber security incidents can be handled seamlessly and do not introduce unnecessary waste to value stream.
- *Acquisition, development and maintenance of information systems*—creation of environments, deployment of codebase, integration testing with different types of hardware devices, complex user acceptance testing tend to result in complex manual work. Reducing manual work to minimum will improve the flow through value stream, moreover it will reduce error-prone activities and keep the security measure applied consistently.

3.2 Technical Measures

The following section describes the main technical security measures that should be considered during the implementation of the security measures. It also tries to explain how the particular security measure is affecting various security goals on multiple levels.

3.2.1 Authentication and Authorisation

Each asset which interacts with, e.g., a user or an external service or is a part of IoT ecosystem such as software components, cloud services, smart devices and hardware components needs to have its own identity. The identity of the asset is subject to authentication (prove that asset is the one) and authorisation (the identity has access to requested data or service). In general we can distinguish three types of identities [12]:

- *Social identity* came up with social rise of the social networks and this type of identity is built by social interaction of the asset (in this case user of our system). The social identity brings the value to the data processing layers of the IoT ecosystems.
- *Concrete identity* provides relation to real 'physical' asset, user or service and needs to be used where relationship between asset and identity cannot be vague. Subscriptions to ecosystem services, device registration, data sharing approvals require more reliable relationships. The relationship can be built by user's email, credit card, phone number, etc.
- *Thin identity* (e.g., username and password) does not have any physical asset relationship.

Example For example, in the VICINITY Platform the *social identity* is applied in the layer of value-added service (Fig. 7.2), where the data is subject to various aggregations, machine learning and artificial intelligence algorithms. The *concrete identity* is required in the VICINITY Cloud services where device and service registration and sharing approvals are required. The *thin identities* are assigned to

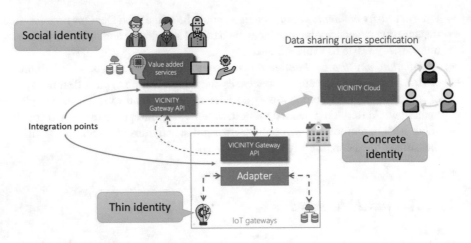

Fig. 7.2 VICINITY identity architecture

devices, services and access points to access peer-to-peer network. These identities are linked to concrete ones on the level of the VICINITY Cloud.

Authentication Mechanisms

As it is previously discussed, each IoT ecosystem includes different types of identities. Even in the identity type there can be several technical solutions for identity representations. Moreover, IoT ecosystem can include large amounts of connected identities communicating with each other [16]. For authentication mechanism it is important to consider the following properties [18]:

- *Scalability*—verification of the identity is the most used service in the whole IoT ecosystem, considering an expected large number of identities being verified. From practical implementation point of view it is worth to focus on external authentication services if possible.
- *Interoperability*—different types of IoT communication protocols provide different types of identities. Hence, the interoperability is critical between these protocols. Identity federated solution is possible [18], however, from the evolution and competition between protocols the interoperability scissors are getting wider. For example, OCPP v1.6 protocol for charging added in version 2.0 has several authentication options. These authentication mechanisms are driven by requirements of eVehicle with weak support of cross domain interoperability [1].
- *Mobility*—smart devices are by nature nomadic, thus their location can change any time. This property is important when the smart devices are connected through the IoT gateways. Some gateway based protocols can handle this use case seamlessly [3]. However, in a situation when a smart device is connected through BLE to a Smart Phone and the Smart Phone is connected to IoT Platform, the

authentication needs to be broken into several services to be inline with identities aggregation and grouping.

Authorisation Mechanisms

When the identity is verified through the authentication mechanism, it is necessary to decide whether requested data and service will be provided or executed. The decision is made by the authorisation mechanism. From the internet of things nature point of view three different authorisation principles can be recognised [17]:

- *Cloud Computing Layer*: the traditional approach from client-server architecture, where the whole authorisation mechanism is centralised on the cloud computing layer. The authorisation policies are specified and enforced in this layer.
- *Fog Computing Layer*: the fog computing nodes enforce authorisation policies for domain, IoT infrastructure or ecosystem that is under their control.
- *Edge Computing Layer*: edge's device enforces authorisation rules on the service they are executing.

Example One example based on the VICINITY platform is shown in Fig. 7.3.

The *cloud computing layer* is used for specifying device and service sharing rules. Each integrated IoT infrastructure, ecosystem or value-added service has its own specific authorisation services, these are kept in fog computing layer and are isolated from the cloud computing one (in this example VICINITY Gateway API components). The edge computing layer is isolated behind the fog computing node which results in a cascade authorisation mechanism (Fig. 7.3).

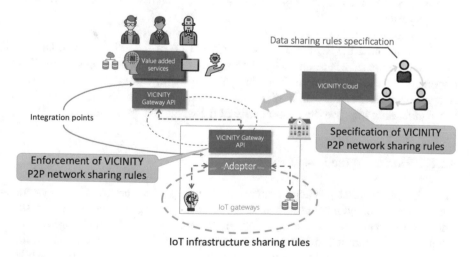

Fig. 7.3 VICINITY authorisation architecture

3.2.2 Passwords and Security Keys Storage

Passwords and security keys are the valuable assets which need to be protected so that users and devices can connect to IoT platforms securely. Implementation of constraints on information entropy of passwords and security keys is crucial to withstand rainbow and dictionary attacks. However, the question that could be asked here: *'How to securely store the password in IoT application in case the storage is stolen?'* The *password and security keys salting and peppering* is the industry standard of protecting stored passwords and security keys [12]. Salts can be stored side by side with hashed passwords, however, pepper has to be stored separately in a single location. Salting and peppering with high information entropy and regular changes of passwords and security keys may improve overall security of authentication mechanism. However, these measures are unable to protect from phishing and social engineering, malware and brute-force attacks.

3.2.3 Data at Rest

Passwords and security keys are the most valuable information stored in IoT platform and devices. But, the loss of stored business relevant data can cause catastrophic scenarios. Data stored in data storage, i.e., Data at rest can be protected by the encryption provided by storage service. It is important to note that encryption of the data at rest can increase the latency in data retrieval, thus it is recommended to test the impact of this measure on overall system performance. It is highly recommended to use strong algorithms in encryption methods. During application of the storage encryption it is important to store the following separately: encrypted data, access control database (e.g., storage administration user login and passwords) and encryption keys.

Note that in some storage engines *Data at rest* encryption needs special license which can increase the total cost of ownership (TCO) of the IoT platform. During the identification of where and what data are stored in the whole ecosystem, it is important to assess the whole data life-cycle thorough, from device to end-user. Encryption of data at rest with minimal set of data stored on devices and in services can significantly improve the data security of the whole IoT ecosystem.

3.2.4 Principle of the Least Privilege

The principle of least privilege requires that an individual program or a system process is not granted more access privileges than are necessary to perform the task. This has implications in the design and implementation of an actual application, setup and configuration of infrastructure and in organisation measures as well. Application user/system roles should be selected and assigned carefully while ensuring that users do not have more rights than necessary.

On the level of the infrastructure it is best practice to have as granular access control as possible and efficient. Different accounts access different data or run different processes. Moreover, in case the 3rd party devices are able to join the IoT platform, principle of the least privilege needs to be extended to these assets, as well. On the organisation level, taking into account the size of the organisation, software developers are not able to deploy software after successful user acceptance testing (UAT) or approve the results of the test.

3.2.5 Data in Motion: Data Transmission Security

Data in motion—data exchanged between parties needs to be protected during its transmission. The most used technique is to protect communication between transport and application layer of the ISO/OSI model using SSL/TLS protocol. The protocol ensures point to-point protection of the communication channel between parties using a SSL certificate. The SSL certificate can be issued by the certification authority (publicly accessible communication) or can be self-signed (testing environment or publicly disclosed production environments). The SSL protocol protects only the communication channel; however, the data transmitted is not encrypted. If data needs to be transmitted between parties through some-kind of a 'broker', e.g., data transmitted from a Bluetooth device to a server through connected smartphone, data should be encrypted and signed. For this purpose asynchronous or synchronous cryptography services can be used [12]. In this case the most critical point is the process of distribution or exchange of public, private or shared keys between parties—it constraints the quality of the security measure.

3.2.6 Infrastructure Level Security

IoT platforms and ecosystems are often a mixture of different technical infrastructures: device to gateways infrastructure, gateway to cloud infrastructure and cloud infrastructure itself. In the device to gateways infrastructure the most critical security goal is to protect pure data collected from the network device and provided to the gateway [13]. They are usually operating in physically constrained environment (e.g., household, office, building) and thus the authenticated encryption feature is the most critical one which is well addressed in most currently used protocols such as Z-Wave, ZigBee, KNX and EnOcean.

The gateway to cloud infrastructure ensures that data is transported through a publicly reachable environment. Hence, more security services are required in this layer. The most notable MQTT, LoRa, Sigfox and CoAP have well established security best practices, however, improvements in preventing DDoS attack are the most critical ones [11].

The cloud infrastructure is the most developed from the security services point of view. It defines the best practices in firewalls setups, intrusion detection and intrusion prevention systems. Fortunately, cloud services often include an IDS/IPS protection, and although it cannot be said that every provider runs these protection services, a provider that demonstrably runs them, can be chosen from a vast number of options. During selection of the infrastructure it is recommended to check how the protocol reacts on threats such as: man in the middle attacks, physical attacks, rogue gateway attacks and routing attacks, and how these attacks impact the IoT ecosystem in overall [3].

3.2.7 High-Availability Deployments

Internet of Things is a domain with intensive data collecting, storing and processing. Applying traditional client–server architectures principles, where high-availability techniques and best practice are well established, can reach its boundaries, especially from total cost of ownership point of view, i.e., processing, storing all data in one place can ramp-up the price of cloud infrastructure bill. The data storage and processing need to be pushed down value chain of the IoT ecosystem towards distribution/decentralisation of the ecosystem, fog and edge computing architectures. Right mixture of these architectures can deal with issues, such as singular resource consumption intensive points, data throughput bottle necks and single point of failure.

Example In the VICINITY ecosystem value-added services are the ones with data processing intensive tasks. The value-added services together with smart devices can be considered as edges of the VICINITY IoT ecosystem. The edges are grouped under one or several fog computing nodes equivalent to the VICINITY Nodes as shown in Fig. 7.3. Each VICINITY node includes three components: VICINITY Gateway API (responsible for authorisation and data exchange in VICINITY P2P Network), VICINITY Adapter and Agent (responsible for providing access to the edges). The only central component in the architecture is the VICINITY Cloud which provides only *device and service registration* functionality and authorisation services. No data coming from the edges is stored in the cloud.

4 Conclusions

The security is the domain which is usually considered as a pure technical one. However, looking at the process of building security framework it is obvious that roots of the security come from the business processes. Without understanding what are security goals of IoT ecosystem stakeholders, the security analysis tends to be out of focus and it will introduce inefficiencies into the IoT ecosystem for users and operations. Starting looking at the security from business perspective will enable us to identify relevant security risks, prioritise them and identify security measures required to mitigate them. It is important to be aware that security risks are not permanently resolved, only mitigated to an acceptable level. When security risks are identified, dealing with security measures especially technical ones is the last step in the process of implementing the security framework. This step is of high importance for achieving secure IoT systems. The selected concepts of the technical measures should help during the selection of the protocols and technologies for the IoT ecosystem. However, complexity of IoT ecosystems results in multi-layer architectures with different mixtures of the protocols and technologies which brings up the interoperability issues in the security domain—the one of the key research challenges.

References

1. Alliance, O. C. https://www.openchargealliance.org/protocols/ocpp-20/
2. Bauer, M., Boussard, M., Bui, N., Loof, J. D., Magerkurth, C., Meissner, S., et al. (2013). *IoT Reference Architecture. Enabling Things to Talk* (pp. 163–211). https://doi.org/10.1007/978-3-642-40403-0_8
3. Butun, I., Pereira, N., & Gidlund, M. (2018). Security risk analysis of LoRaWAN and future directions. *Future Internet, 11*(1), 3. https://doi.org/10.3390/fi11010003. https://www.mdpi.com/1999-5903/11/1/3/pdf
4. Dabbagh, M., & Rayes, A. (2018). Internet of things security and privacy. In *Internet of things from hype to reality* (pp. 211–238). https://doi.org/10.1007/978-3-319-99516-8_8
5. Foundation, T. O. https://www.owasp.org/index.php/OWASP_Internet_of_Things_Project#tab=IoT_Vulnerabilities
6. Foundation, T. O. https://www.owasp.org/index.php/OWASP_Internet_of_Things_Project#tab=IoT_Attack_Surface_Areas
7. Ghirardello, K., Maple, C., Ng, D., & Kearney, P. (2018). Cyber security of smart homes: Development of a reference architecture for attack surface analysis. In *Living in the Internet of Things: Cybersecurity of the IoT - 2018.* https://doi.org/10.1049/cp.2018.0045
8. Gruschka, N., & Jensen, M. (2010). Attack surfaces: A taxonomy for attacks on cloud services. In *2010 IEEE 3rd International Conference on Cloud Computing.* https://doi.org/10.1109/cloud.2010.23
9. Hackenberg, R., Weiss, N., Renner, S., & Pozzobon, E. (2017). Extending vehicle attack surface through smart devices. *The Eleventh International Conference on Emerging Security Information, Systems and Technologies*, ThinkMind, 131–135.
10. International Organization for Standardization Geneva, Switzerland (2011). ISO/IEC 27005:2011 Information technology – Security techniques – Information security risk management. https://www.iso.org/standard/56742.html
11. Kim, J.Y., Holz, R., Hu, W., & Jha, S. (2017). Automated analysis of secure Internet of things protocols. In *Proceedings of the 33rd Annual Computer Security Applications Conference on - ACSAC 2017.* https://doi.org/10.1145/3134600.3134624
12. Leblanc, J., & Messerschmidt, T. (2016). Identity and data security for web development: Best practices. Sebastopol: O'reilly Media.
13. Marksteiner, S., Jimenez, V. J. E., Valiant, H., & Zeiner, H. (2017). An overview of wireless IoT protocol security in the smart home domain. In *2017 Internet of Things Business Models, Users, and Networks.* https://doi.org/10.1109/ctte.2017.8260940
14. Reshetova, E., & McCool, M. (2019). *Web of Things (WoT) Security and Privacy Considerations.* https://www.w3.org/TR/wot-security/#wot-threat-model-stakeholders
15. Soós, G., Kozma, D., Janky, F. N., & Varga, P. (2018). IoT device lifecycle—A generic model and a use case for cellular mobile networks. In: *2018 IEEE 6th International Conference on Future Internet of Things and Cloud (FiCloud)* (pp. 176–183). Piscataway: IEEE.

16. Thakkar, D. (2016). *Internet of Things and Multifactor Authentication*. https://www.bayometric.com/internet-things-multifactor-authentication/
17. Xu, R., Chen, Y., Blasch, E., & Chen, G. (2018). *A Federated Capability-Based Access Control Mechanism for Internet of Things (IoTs)*. https://arxiv.org/pdf/1805.00825.pdf
18. Zhu, X., & Badr, Y.: Identity Management Systems for the Internet of Things: A Survey Towards Blockchain Solutions. Archives-ouvertes.fr (2018). https://hal.archives-ouvertes.fr/hal-01945947

Chapter 8
Privacy, GDPR, and Homomorphic Encryption

Christopher Heinz, Nigel Wall, Alexander H. Wansch, and Christoph Grimm

1 Challenges to Privacy from the IoT

When the United Nations proclaimed The Universal Declaration of Human Rights (UDHR) [25] at the UN General Assembly in Paris on 10 December 1948 it was a milestone document in the history of human rights. Drafted by representatives with different legal and cultural backgrounds from all regions of the world it continues to be seen as the standard to be achieved for all peoples and all nations. It set out, for the first time, fundamental human rights to be universally protected. At that time they could have had little foresight into how computers would become ubiquitous, with massive deployment of sensors and the creation of big data resources.

Article 12 of the Universal Declaration of Human Rights (UDHR) states: "No one shall be subjected to arbitrary interference with his *privacy*, family, home or correspondence, nor to attacks upon his honour and reputation. Everyone has the right to the protection of the law against such interference or attacks" [25]. That objective remains equally valid today, but now there are so many new ways that privacy can be infringed. Indeed, the UDHR has been further developed in Europe as the CHARTER OF FUNDAMENTAL RIGHTS OF THE EUROPEAN UNION (2012/C 326/02) [13], which includes Article 7: Respect for private and family life: "Everyone has the right to respect for his or her private and family life, home and communications" [13].

C. Heinz (✉) · A. H. Wansch · C. Grimm
TU Kaiserslautern, Kaiserslautern, Germany
e-mail: heinz@cs.uni-kl.de; a_wansch14@cs.uni-kl.de; grimm@cs.uni-kl.de

N. Wall
Climate Associates Limited, Ipswich, UK
e-mail: nw@nigel-wall.co.uk

© Springer Nature Switzerland AG 2021
C. Zivkovic et al. (eds.), *IoT Platforms, Use Cases, Privacy, and Business Models*,
https://doi.org/10.1007/978-3-030-45316-9_8

Internet of Things (IoT) objects often operate in a personal, private environment. Devices such as smart watches, fitness trackers, and medical devices being some obvious ones. Smart home devices such as motion and temperature sensors or even smart speakers such as Amazon Alexa and Google Home extend the range of IoT objects. There are many more connected sensors the data subjects may be less aware of, all of which are generating, processing, and transmitting personal data to potentially untrusted and potentially malicious cloud services. User privacy remains an important issue and one of the key acceptance criteria when developing and deploying IoT Systems [18].

It is essential to adopt the principle of "privacy by design" and "privacy by default". Privacy by design as a concept was introduced in 1995 and led to the identification of seven fundamental principles [5], which have been developed into evolving legislation, with the latest and most comprehensive legislation being the Europe-wide General Data Protection Regulation (GDPR) [10].

Regulation requires privacy to be protected by restricting the use that can be made of personal data, and the first question is to define what data sets will be considered to be Personal Data. Clarification over specific questions relating to privacy were considered by the Article 29 Data Privacy Working Party which was replaced by the European Data Protection Board (EDPB) [12] on the day that the GDPR became law. However, the Article 29 WP archives [11] make interesting reading.

Essentially, if there is a data set that contains personal information about an individual which does not in any way identify the individual that it relates to, then it ceases to be personal data. However, if it is possible to link this data set to the individual by applying another source of knowledge, then the data set must be considered to be personal data. An example might be information about an unidentified car that made a journey from a private residence to an office block at 8:00 AM: it would be possible to work out exactly who lives at the residential address and works at the destination office. In which case the information about that journey would have to be considered to be personal data.

The easiest way for an organisation to fully protect personal information is not to handle personal information at all, and certainly not to store it. If personal information has to be processed, then the risk of unauthorised access to the personal data is limited by deleting any such data as soon as it is no longer needed.

It is important to understand the difference between privacy and security. The legal requirement is to protect privacy. One of the threats to privacy is from cyber-attack. Whilst the provision of excellent security is a critical requirement to protect privacy, it is not sufficient on its own. Privacy could be breached where encrypted data is passed securely to a database, but then the authenticated Data Processor carries out actions on the data that is for purposes that have not been agreed with the data subject. "Privacy by design" includes the need to protect the personal data from attack, even when the security systems have been breached.

There are three operational domains that need to be considered:

- *IT systems*: as engineers we tend to focus on getting the technology right, with strong encryption and protected databases.
- *Operational methods*: it is essential to design systems that minimise the use of personal data, and the number of places where the data are stored whilst restricting access to those who genuinely need to know the data subject's identity.
- *Physical arrangements*: unfortunately, many security and privacy breaches are caused by IT systems and processes not being used as intended. For example, a USB memory stick may be inserted into a processor that contains sensitive data, and data copied when this has not been authorised.

Some practical steps to protect personal data and satisfy the needs of GDPR rules when developing a new system or service implementation are explained in Sect. 3.

2   General Data Protection Regulations

In this section, we step back from the technical view and review laws and regulations. There is a discrepancy between what is technically possible and what is legally allowed. As already mentioned, the current legal framework provided by the EU is *The General Data Protection Regulation (GDPR)*. The following text focuses on the changes in data protection law compared to former laws and also gives some insight into the practical steps that should be taken when creating a new service that uses personal data.

2.1   Introduction

Regulations concerning data protection in the European Union have evolved slowly. The member states have tended to have national laws and even different laws in their local regions. The regulations immediately prior to GDPR were the European Commission's directive DIR95 from 1995, which was complemented by the Directive on Privacy and Electronic Communications (EC Directive 2002/58/EC) [24]. Since this directive was made the ability and extent of automated data processing far exceeded what was intended to be regulated. Therefore, in 2012, the European Union decided on the development of the new *General Data Protection Regulation* in order to deal with the problems that have evolved due to too weak and non-aligned regulations.

The very same regulation has to be applied in different scales—for a single individual using a personal computer at home, as well as for a global player and multi-billion high-tech company like Facebook. Consequently, it is a very comprehensive legal document with a huge impact on EU-citizens and companies. It also applies to everybody outside of the EU, that handles personal data from inside the European Union. Furthermore, the fines for breaches have increased significantly and this has motivated awareness and the attention of lots of people and legal entities have taken steps to be compliant to the GDPR.

2.2 *Evolution of Privacy Laws to GDPR*

2.2.1 General Provisions, Principles, and Definitions

The GDPR introduces worldwide obligations: the law applies wherever personal data relating to an EU citizen is processed. To help understand the legal requirements new definitions were introduced.

One of them is pseudonymisation, which means that processed data is handled (processed and physically stored) in a way, that it cannot easily be related to the person it belongs to, without additional information. This is different from anonymisation where the identity of the personal that this relates to is removed in such a way that the link cannot be reinstated.

Whilst pseudonymisation reduces the ease with which personal data can be accessed, it does not prevent the decoding of the data. Whilst a single breach of security would be unlikely to result in any loss of personal data, without access to the pseudonymisation tables, pseudonymised data has still to be considered to be personal data, so it should be carefully protected.

New definitions regarding the type of data have also been introduced.

- *Personal data* is any information that can be associated with and identify an individual. This clearly includes any data where the individual is named explicitly, but it also includes any data which can be associated with an individual by correlation with another database. For example: information about a privately owned car is deemed to be Personal Data. Similarly, information from sensors installed in a private home would be deemed to relate to the home-owner.
- *Sensitive personal data*: Much of the data associated with an individual would not be considered to be sensitive personal data. However, information about health, membership of organisations, personal relationships, sexual orientation, religion, criminal activity, and convictions, etc., would be considered to be sensitive because much more damage would result from a data breach. For the purpose of normal operation this is all treated as personal data.

The main practical differences are in the need for greater limitation of processing sensitive personal data and in the way that a data breach is handled and reported. Sensitive personal data also includes: "Genetic data is defined as any data relating to an individual's characteristics that are inherited or acquired during early prenatal development. Biometric data denotes any data relating to an individuals' physical, physiological, or behavioural characteristics and allowing a unique identification. Data concerning health means any information related to an individuals' physical or mental health or the provision of health services to the individual" [4].

Medical data is already strictly confidential and must be kept secured. However, data like an individual's body mass index (BMI) together with their name and/or date of birth becomes classified data. Even if the individual's height and weight are written on a piece of paper, one could calculate the BMI and thus this would be a data breach. Thinking of fitness clubs that provide plans for their members training, this implies a significant challenge that will inevitably result in changes of practice. New principles that can be found in the regulation are, for example, the transparency of data processing which grants the data subject access to information on how their data is processed. In practice this means that any process involving their personal data needs to be explained to a customer in the case he or she has any concerns about the process.

Additionally, the new GDPR clarifies several points that can also be found in the Directive 95. These are in particular:

- *Data minimisation principle* stating that the amount of data kept must be limited to the amount necessary to fulfil the task for which one has given the consent.
- *Processing of children's data.* If children are under 16 years of age, processing of data requires the consent and authorisation of a parent or custodian. This implicitly limits any use of online content which is intended to be for minors, especially under the age of 16. So far it appears that providers of such content may need to put much effort on implementing this restriction.

2.2.2 Transparency and Modalities

This category is closely related to the principle of transparency of data processing described above. It forces a data controller to determine the purposes of the processing of data. There must be an Explanation of conditions for the consent. The consent must be given freely and explicitly, can be withdrawn at any time, and must not be expanded to any corresponding process without prior permission.

2.2.3 Information and Access to Personal Data

This category expands the data subject's rights to get information from the data controller about transfers of data to third countries. It also requires a mechanism for the data held to be accessible to the data subject (who may then require corrections to be made). Furthermore, information on the processing and purpose of processing must also be provided on request. Information about the data subject's rights must be provided. This point, as well as the previous one, forces data processing entities to be aware of their own processes and to check their lawfulness since they may be required to explain what they are doing.

2.2.4 Rectification and Erasure

There are two very important rules in the GDPR that need to be considered: *right to rectification* and *right to erasure*. The *right to rectification* obliges any data processing or storing entity to correct any data if requested by the data subject, as well as restricting the processing of the data to the amount authorised by the data subject. Additionally, the *right to erasure*, which is also known as the right to be forgotten, grants the data subject the right to instruct the deletion of their data. This erase has to happen in a cascading manner, particularly if the processor or the controller has handed over any data to a third party. These two rights oblige data processing or storing entities to be aware of any relation to any data subject and, as well be aware of the storage location(s) of all data in their possession.

2.2.5 Right to Object and Automated Individual Decision Making

These two points refer back two corresponding articles in the DIR95. First, the *right to object* forces the processor of personal data to provide legitimate reasons for his work and the processing of the data. This must be stated explicitly. The second point protects the data subject from being evaluated and maybe even categorised automatically without human interaction. If consent to automated processing is not given explicitly, then the data subject should be provided with the opportunity to conclude their business through an interaction with a natural person. However, there is a real threat that service providers will simply decline to provide the required service if the data subject is unwilling to allow the use of their personal data.

2.2.6 General Obligations

This category contains at first the principles of *data protection by design* and *by default*. *Data protection by design* implies that, for every new process implemented, data protection must be a major design criterion. This forces companies to be aware of data protection even at an early point and maybe refuse setting up processes due to their non-compliance. *Data protection by default* implies the highest standard of data protection settings must always be the default. In practice, for example, if one applies for a social network, the privacy settings must be the highest possible and any broadcast of personal data must be refused by default. Further points of this category clarify responsibilities for the processing of personal data. Also an obligation to "maintain record activities [. . .] and co-operate with the supervisory authority" [8]. This clearly puts processors into responsibility and forces them to support supervision authorities, which might improve an open and more aware culture of data processing.

2.2.7 Security of Personal Data

The main point of this category is the obligation to report personal data breaches to a supervisor, respectively, a supervisory authority and to the data subject. So, from now on, one must be informed if any information is leaked or anything contrary to the GDPR happens.

2.2.8   Data Protection Impact Assessment and Prior Consultation

Mainly, this is highly related to the principles of *data protection by design* and *by default*. Controllers are obliged to evaluate the data protection impact prior to a process and only execute it if there is a low or no risk as a result. If a high risk is detected, the supervisory authority must be taken into consideration and contacted in order to evaluate the process together and share the responsibility or even cancel the process.

2.2.9   Data Protection Officer (DPO)

A major new concept of the GDPR is the designation of a data protection officer (DPO). There are several preconditions that obligate an entity to designate someone as the DPO. Mainly, they are as follows: If an entity's core business or major activity is the processing of personal data or monitoring data subjects, if special (in particular sensitive) data is processed or if the entity is a public authority. Any designation beyond these conditions is voluntary. The DPO has several tasks like monitoring the processes, supporting the implementation of new processes, analysing existing processes, informing and training the staff, checking on the general compliance and providing advice. Moreover he/she has to perform risk impact analyses, report to the management board, and also to report breaches to the supervisory authority. The DPO need not be a full time role, but the DPO must be selected carefully as someone who can fulfil his or her duty independently without the risk of any negative consequence regarding his or her other responsibilities.

2.2.10   Code of Conduct and Certification

Basically, the new regulation offers ways to approve the compliance to codes of conduct by supervisory authorities and general "certification, mechanisms, seals, and marks" [24] to show that an organisation complies with the regulation. This is to simplify such processes without running the risks associated with self-certification, which limit the accountability of programs.

2.2.11   Transfer of Personal Data to Third Countries or International Organisations

The articles related to this category aim to strictly limit the uncontrolled distribution of personal data, especially to third according to their level of data protection compliance.

2.2.12 Remedies, Liability, and Penalties

The last category is divided into three parts.

- First, it is now clearly specified how a data subject can take action against an entity believed to be infringing the GDPR. Furthermore, it is clarified which entities are allowed to take action in addition to the data subject taking action. This point is a huge progress in the field of customer rights because it is now much easier to demand one's rights.
- Second, from now on, everybody involved in the processing of data is responsible and can be called to account if a data breach occurs. This makes it harder or even impossible for entities to pass on their liability by processing data through a network of processors.
- Last and most important, the fines that can be imposed have risen significantly and are now at a maximum of 20 million Euros or 4% of the total annual global turnover. Therefore, it has become more important for companies to comply to the GDPR in order to avoid massive, if significant breaches occur.

3 Practical Steps to Protect Personal Data

There are many consultancies who can advise organisations and projects on how to comply with GDPR. Whether one uses a consultancy or tackles the job in-house, there are very useful resources available, such as the British Standards Institution (BSI) guide on 20 steps to GDPR compliance (see [4]). A check list was used in the VICINITY project, that was adapted from the BSI guidelines. The steps developed for VICINITY are presented below in chronological order, with later activities needed as the design and implementation moves ahead:

1. There should be a project Ethics Board that will appoint a project DPO (to work with DPOs in each company involved in the collaboration). Appointing a DPO is a legal requirement if it is considered that a breach of personal data privacy could have a major impact. In any case, it is still worth having a DPO.
2. Identify and assign responsibilities to Data Controllers (DC) and Data Processors (DP) for the new value-added services that are to be provided.
3. DPO should create and maintain a register to record the achievement of all the actions identified below, as an aid to managing the process and evidence that good processes were followed—which could be vital evidence of mitigation measures taken, if there is ever a data breach that leads to legal action being taken against the organisation.
4. Training should be provided to people with new responsibilities for protecting personal data. (Company top-team, DPO, DCs, DPs, and Ethics Board as a minimum.) Also, raising awareness among everybody being involved leads to an increased acceptance and cooperation.

5. Data controllers create data/information registers to identify all the types of personal data and sensitive personal data that will be collected. Headings for information to be included in the register include:

   (a) For each system that handles personal data, create a file that identifies the nature of personal data held
   (b) Record why it is held and how it will be processed.
   (c) Explain how will legal consent for this purpose will be obtained
   (d) Define the required data retention period and data destruction date.
   (e) Identify the set of people that will be granted access to this personal data, keeping this to the minimum number possible.

6. Data controller works with the system architect to create a data flow diagram—showing all the subsystems that the data passes to or through.

7. Ensure adequacy and non-excessiveness: review the data flow diagram and identify any steps that can be taken to reduce the number of subsystems that have sight of the personal data and also to reduce the overall collection of data.

   (a) Remove personal identification as soon as possible leading to anonymisation.
   (b) If anonymisation is not possible, use pseudonymisation as soon as possible, keeping the pseudonymisation table in a core database. This will mean that if (when!) a data breach does occur, it will only release pseudonymised data which the unauthorised third party cannot understand unless they can also hack the pseudonymisation table.

8. Try to minimise the risk of wider distribution of personal data by avoiding use of third-party services, using in-house systems by preference. Where third-party suppliers must be used, choose these carefully based on an assessment of the risks of leakages occurring. Also, try to establish a contractual framework w.r.t. compliance.

9. DPOs should oversee the creation of briefing packs including a Privacy Notice and consent forms to confirm permission for current and intended uses of the personal data both, for internal and external stakeholders' information.

   (a) Fully explain what data will be collected, why the data is needed, all the uses that will be made of the personal data, who will have access to personal data, where data will be stored, how long it will be held, who to contact if the data subject has worries or wishes to change the permission that they have granted.
   (b) Explain the rights that the data subject has to: inspect data held; right to rectify; withdraw permission to use; decline profiling; instruct transfer of data to another service provider or to have all records deleted.

10. Data Retention: the data controller should check that personal data has been destroyed by the agreed destruction date.

11. DPO monitors that data subject rights are being observed.

12. Data breach response: Develop and install a plan for what to do when a data breach does occur: who plays what roles in the company, who to inform, handling press enquiries, etc. The legal framework provides duties such as reporting within 72 h to the supervising data protection authority and the data subject.
13. Review what has been implemented. Are there any locations where personal data is held that could be at risk of a physical attack, hack, or leak.
14. Produce a baseline Data Protection Impact Assessment (DPIA)—what would be the result of a failure? This needs to feed into project and company risk registers.

    (a) Note that the production of a DPIA is a legal requirement if the activity is deemed to be "Likely to produce a high risk", in other case it is just good practice.

15. Update the DPIA considering operational aspects.
16. Ensure that the company/project senior management continues to be committed to the actions needed, to ensure personal data are properly protected.

    (a) Management Board should be made aware of and discuss: incidents, near misses, Data Subject requests, DPIA reviews, performance of third parties trusted with personal data

17. Review and revise (if necessary) the company Data Privacy Policy.
18. Communicate to, and continue to train system architects, designers, developers, integrators, and operators on the approach that is to be taken, regarding handling of personal data, and the need to mitigate risks at all stages.

4 Database Protection, Moving Forward from the State of the Art

If personal data has been collected from sensors that cannot be interfered with and passed securely to a database, there remains the very real threat that authentication errors may allow unauthorised users to access the data. Furthermore, authenticated users may decide to use the data for purposes that have not been agreed by the data subjects. So, there needs to be limitations on the nature of the processes that will be allowed. The state of the art in IoT cloud platforms is led by two companies according to [9]:

- Amazon Web Services IoT (AWS) (34% of the 2018 market)
- Microsoft Azure IoT (23% of the 2018 market)

For a broader overview considering a wider range of platforms, please refer to Chap. 2, or see [16].

4.1   Amazon Web Services (AWS)

The most dominant Service for realising IoT Applications currently is the Amazon Web Services (AWS) platform (see [1]). Amazon offers a quite extensive overview and documentation about their AWS platform. In this chapter, we will focus more on the security and, in particular, privacy aspects of the platform, rather than analysing the platform itself. For a more detailed overview and documentation on how to use AWS, interested readers are advised to consult the developers documentation (see [1]).

Data transmission between AWS and respective endpoints is secured using TLS v1.2. Furthermore, both communicating parties (e.g., clients and server backend) are authenticated using state-of-the-art X.509 Certificates. Application of these certificates on the end devices is mandatory as the AWS Message Broker will check and authorise all access before any incoming message is relayed. Vice versa, end- or edge-devices can verify and validate incoming messages using the AWS server's certificate.

Once devices and server are authenticated and communication between them is secured, the question remains: how is privacy addressed in this context?. As the users data is secure inside the AWS platform and while being transmitted to the end devices, the challenge of privacy first of all comes down to authorisation: who is allowed access to what kinds of data and for which purpose? In AWS, data access is controlled via Policies. A fine grained selection and orchestration is possible in order to control and restrict access of the attached and registered services. E.g. specific services can be granted read/write access only to certain Message Queuing Telemetry Transport (MQTT) topics and read data from there. Yet the received data in clear text can be further relayed and forwarded to third-party services outside of the AWS platform and hence outside of the control of the user, data owner, and naturally also AWS itself.

4.2   Microsoft Azure

Microsoft Azure (see [19]) follows quite a similar approach, when it comes to security and privacy. Again, Microsoft has documented their platform quite well and interested readers are pointed to the developer documentation to get a more detailed insight into the platform and its architecture (see [19]).

As with AWS, also on Microsoft Azure, any communication between the backend and the end devices is secured using TLS v1.2. User end devices and services are again also authenticated using X.509 Certificates. Hence data inside Microsoft Azure can again be considered secure. A brief look also reveals similar insights on privacy as was the case with Amazon Web Services: Developers can set up rules to forward incoming data/messages in clear text to other Azure Services and hence, e.g., be stored in a database.

5   Next Step: Homomorphic Encryption?

Currently top cloud providers, described above, do their best to ensure data
protection of their users' data. Yet at some point this data will need processing
by a third-party service. The aforementioned providers offer means to control and
authenticate *WHO* is able to access private data and to protect it from unauthorised
access in terms of security as best as possible. Yet there is no possibility to control
HOW this data is processed and *WHAT* is being done with it other than trust and
legal claims. And yet at some point, service providers are to be trusted. Some may
be malicious: it can be very difficult for the data subject to know who to trust. This
typically leaves the user with two possible options. He can either:

- blindly share his private data with others in order to continue using their services,
  or
- only share his information with people and organisations that are fully trusted,
  which drastically reduces the benefits we can gain from the emergence of the IoT.

A third option which will combine exists the benefits of both options listed above.
This is the use of homomorphic encryption that allows data sharing with partially
trusted third parties whilst restricting the operations that they can perform on the
data, so keeping the shared data protected.

5.1   What Is Homomorphic Encryption?

Homomorphic encryption (HE) uses a public-key encryption scheme with homomorphic properties. This process relies on the ability to carry out the required
operations on encrypted data with the result being equivalent to this operation being
performed on the clear-text data first and the result being encrypted afterwards:

$$Enc(A) \circ Enc(B) = Enc(A \circ B)$$

This approach means that neither the source data nor the resultant data can be
understood or decoded by the system that is carrying out the operation.

5.1.1   Rivest, Shamir, and Adleman Encryption (RSA)

The idea of homomorphic encryption is as old as public-key encryption. Homomorphic properties are given in the RSA encryption, which was introduced in 1978 by
Rivest, Shamir, and Adleman [21]. Using the RSA encryption scheme, two cipher
texts c_1 and c_2, given as

$$c_i = enc(m_i) \equiv m_i^e \mod N$$

can be multiplied by

$$enc(m_1) \cdot enc(m_2) = (m_1^e \cdot m_2^e) \mod N = (m_1 \cdot m_2)^e \mod N = enc(m_1 \cdot m_2)$$

However, this representation of homomorphic encryption is really simplified. The basis are comprehensive number-theoretical considerations about ideals, lattices, and ideal lattices. This scheme shows the basic functionality of homomorphic encryption. For a further reading on the theory below, we refer a reader to the mathematical works of Paillier (see [20]) and Gentry (see [14]). Also several improvements by Vaikuntanathan, Brakerski, van Dijk, and Halevi (see [2, 3, 7, 15]) are subject to a further reading.

As the plain RSA or "Textbook-RSA", which was outlined above, is not considered secure anymore, a padding mechanism was introduced to avoid possible attack patterns. However, these padding functions break the homomorphic properties of RSA. Additionally, one might require not multiplication, but rather the addition of c_1 and c_2.

5.1.2 Paillier Encryption

In 1999, Pascal Paillier published his work on a new "Public-Key Cryptosystems Based on Composite Degree Residuosity Classes" [20]. Paillier's encryption scheme allows for the two ciphers c_1 and c_2 to be added up but not multiplied. This operation is much more common and will be required for the data protection approach, which is further described in Sect. 5.2.

We want to give a short insight on the basics of the considerations by Paillier. He first introduces Composite Residuosity and classes of numbers that fulfil this property. The mathematical representation is as follows: with a given number $n = p \cdot q$ and p and q being two large primes, a number z is the n-th residue modulo n^2 if a special y exists. Then z is defined as:

$$z = y^n \mod n^2$$

As a link to RSA (see Sect. 5.1.1), Paillier elaborates the hardness of finding the n-th modulo n^2 residue and therefore states the intractability of Composite Residuosity. This results (with several steps in between) in an encryption/decryption scheme that looks as follows:

Encryption plaintext $m < n$, a randomly selected base g and a random number r. (n, g) is public but (p, q) is private.

Then the ciphertext c is derived as:

$$c = g^m \cdot r^n \mod n^2$$

The decryption works as follows:

$$m = \frac{L(c^\lambda \mod n^2)}{L(g^\lambda \mod n^2)} \mod n$$

For an elaborate explanation of the functions L and λ, we refer the reader to the original work [20], since this lies out of the scope of this chapter.

Finally, Paillier shows that the property of homomorphic additivity holds for the encryption function. Therefore, we obtain another step into the direction of (fully) homomorphic encryption.

5.1.3 Fully Homomorphic Encryption by Craig Gentry

Each of the above schemes allow only a limited number of operations (i.e., only multiplication or only addition but never both). This is often referred to as "partially homomorphic encryption" (PHE). New approaches to fully homomorphic encryption were made, based on the mathematical principle of ideal lattices. In a first step a *somewhat homomorphic encryption* scheme was designed. The problem being some noise introduced to the encrypted cipher text, which was adding up with each homomorphic multiplication, increasing the cipher text size and at some point even making decryption impossible.

Finally, some schemes also allowed arbitrary circuits or functions being computed on homomorphically encrypted cipher texts [22]. Yet still, these cipher texts were increasing in size exponentially to the depth of the circuit evaluated, making it unfeasible for repetitive evaluations of cipher texts.

However, in 2009, Craig Gentry published his PhD thesis [14] introducing "A fully homomorphic encryption scheme" by applying bootstrapping on ideal lattice based, *somewhat homomorphic encryption* schemes. With the results of Gentry's dissertation, it is possible to evaluate arbitrary functions on cipher texts, however, at the cost of computational demand and key size.

A more in-depth evaluation on the impact and effect that using homomorphic encryption has on the overall runtime was conducted in [17].

5.2 How Is Data Protected with Homomorphic Encryption?

We can utilise the homomorphic property of certain cryptosystems to allow for secure sharing of private data with potentially malicious or untrusted third parties. One example could be a value-added service offered by someone else. If a user is expecting some particular benefits of the value-added service or is even forced into using it, keeping his sensitive data private is a major concern. Missing trust can potentially even be a showstopper for the whole Internet of Things.

We propose the following approach, for secure data sharing: First and foremost, only encrypted data should be shared. Encrypted user data could be collected and stored in the cloud, so that storage and requests are performed in the cloud on a more powerful device. The basics on how a secure end-to-end communication can be established are presented in Chap. 7.

This is an important mechanism against eavesdropping. Yet it does not solve any privacy concerns in a scenario, where the receiving party is malicious, compromised or for other reasons simply cannot be trusted. As explained in Chap. 7, security and privacy always come down to a matter of trust. Who can be trusted? Do I really need to give out my sensitive data to this service? Do I need to give out any sensitive data at all? In many cases, we claim that the latter is not necessarily required.

An example utilising homomorphic encryption in combination with a fitness-tracking app is described in [23]: A fitness-tracker uploads encrypted fitness data onto the cloud and later asking for "average heart rate over the last 24 hours": in this case, decryption is done at the user's site, with no clear text visible to third party at all. This enables moving data storage and parts of the data processing onto a more powerful cloud server without disclosing any clear text information to said server, as decryption is done at the user's site. If the user keeps his/her decryption key on, e.g., his laptop and his smartphone, he is able to decrypt the data on both devices and can hence still take advantage of the benefits a cloud offers without trusting it.

In a multi-party or data sharing case, encrypted data is again uploaded to a VAS. Only this time the data is aggregated using the inputs of multiple parties. Only when enough inputs have been collected, the encrypted data is immediately aggregated and then all the personal data deleted. With a sufficient number of sources the aggregated data can be considered to be anonymised—anonymised/aggregated data can be decrypted and made available to VAS for further processing outside the constraints of GDPR.

A typical example for such a scenario would be an energy demand management. Participating households are required to send their current or—in an even smarter environment—planned energy consumption to a common grid operator. The operator is in generally only interested in the overall consumption of a whole district or city, not the individual households. Yet, as energy consumption can lead to conclusions about personal behaviour or even simpler tell whether a user is at home or not, this as well has to be seen as private data, which should not be revealed. Using homomorphic encryption, we can upload encrypted data only and aggregate (e.g., sum up) all reported energy consumption in encrypted form. The aggregated result can finally be decrypted with the consent of all participating parties. A more in-depth protocol on how this kind of secure multi-party computation can be achieved utilising homomorphic encryption is given by Damgård et al. [6].

The use of the full capability HE processing will not be required in all cases. In some cases PHE will be sufficient such that a service only requires an aggregation (e.g., the sum) to operate and will only collect the provided inputs just to calculate the aggregated value.

5.3 Homomorphic Encryption and GDPR

GDPR places a number of requirements on the Data Controller and the Data Processor. The main requirements are that personal data must be kept secure, accurate, and private, that it must only be used for the stated and agreed purposes, that the Data Subject can examine data records and insist on errors being corrected. If the Data Subject wishes all their personal data records must be transferable from one organisation to another, and ultimately the Data Subject can insist that all their records are deleted.

It must be recognised that encrypted data that are shared remain personal data, so there are constraints on where the data may be stored. Specific permission would be required to allow the personal data to be aggregated for an agreed purpose. If the encrypted personal data could ever be stolen and decrypted, then there would be a major breach of the regulations, with consequential fines. Once the data elements have been aggregated with a large number of other data elements, then the resultant aggregated data set would not be considered to be personal data and would be outside GDPR. It would be wise to ensure that as soon as the required aggregation had been completed all the encrypted personal data should be deleted. No copies of the encrypted personal data should be kept (except perhaps short-term back-up which would be deleted as the working files were deleted). The danger with holding files of personal data that are strongly encrypted by today's standards, within a few years stronger code cracking/decrypting schemes will be introduced and these data would no longer be secure.

The practical implication is that a third party could operate a database with encrypted personal data held in limited-time storage, such that "untrusted" organisations would be allowed to perform analysis of the data (to the extent and for the purposes permitted by the Data Subjects) and would be able to receive the results of their analysis without ever seeing any source data. This still leaves the challenge of how to ensure that the data analysis carried out falls within the range of usage that the Data Subjects have agreed. So, in practice all users would need to be trusted, but the process described here does reduce the attack surface and make the whole process more secure.

5.4 Example: VICINITY IoT Platform

The IoT platform implemented within the EU funded project VICINITY was designed following the principle of "privacy by design". Hence, a basic idea in the project was to disclose the fewest amount of data needed. In order to do so, a user must first actively give his consent to share data and with whom to share it. Any service, that uses data from users, has to be authorised before. This authorisation includes consent to the explicitly described terms and conditions of a service and can be withdrawn at any time. As VICINITY itself only provides an infrastructure

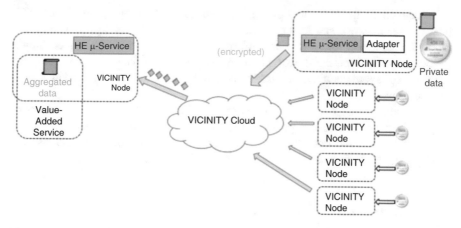

Fig. 8.1 Homomorphic encryption micro-service applied to use case

for peer-to-peer communication, afterwards, user data is only transmitted from the data subject directly to the consented value-added service (VAS) and never stored on any cloud on its way.

Now we take homomorphic encryption and aggregation into consideration. Once we have data that has been encrypted homomorphically, an aggregation is performed on the resulting data. The result now is anonymised and outside the scope of the GDPR.

The VICINITY project has investigated the use of homomorphic encryption for the above scenarios. As shown in Fig. 8.1, Energy Consumption Data of some networked devices is gathered and transmitted to the operators' value-added service. The operator is only interested in the overall energy consumption and so it calculates the sum over all inputs first. As the energy consumption allows to draw some conclusion about the owners behaviour, we will consider this data as private data and also wish to keep this data private and not share it with any third party.

As neither the operator nor the users themselves have any particular interest in sharing their individual, private data, we introduce the homomorphic encryption micro-service. It is integrated into the VICINITY dataflow as shown in Fig. 8.1: Input Data (e.g., the Energy Consumption) is encrypted using a homomorphic encryption scheme. In the case of partially homomorphic encryption, we used a simple Paillier encryption (see above in Sect. 5.1.2, [20]), which allows addition of cipher texts and thus to calculate the overall energy consumption of all attached devices. Hence, PHE is sufficient for this purpose.

In other use cases, addition might not suffice. Hence, we further investigate the use of Brakerski–Gentry–Vaikuntanathan (BGV) scheme, which is a fully homomorphic encryption scheme implementing performance improvements over the original fully homomorphic encryption by Craig Gentry (see above in Sect. 5.1.3, [2]).

In both cases, the encrypted payload is then transmitted through the VICINITY peer-to-peer network and to the operators value-added service again. The homomorphic encryption micro-service performs the addition on the encrypted inputs, followed by a decentralised decryption on the aggregated data. This way, only the anonymised, aggregated data is visible and handed over to the value-added service in the first place.

One concern that remains is that homomorphic processing requires orders of magnitude more processing power, which is now feasible but does introduce a processing delay—so not ideal for real time processing—and does increase the energy used by these systems.

6   Conclusion

In this chapter, we discussed the legal obligations with respect to peoples' privacy. Apart from obvious personal doubts and issues raised, since 25th May 2018 every personal data being processed in the European Union underlies the General Data Protection Regulations or GDPR for short. Due to globalisation, especially in the IT sector, this has worldwide effects as any person or company in business around the EU is obligated to comply to these regulations. We have outlined the most dominant changes and regulations and have given a step-by-step guideline following the recommendations of the British Standards Institution (BSI) [4] in order to comply with the GDPR.

After the legal framework for privacy was discussed, we have given examples in the field of IoT, which are most commonly used nowadays. Most of the time, these frameworks offer an infrastructure or means of establishing communication between attached devices (machine to machine) or between devices and services. Following latest standards and recommendations, these frameworks offer a very sophisticated approach to security (see Chap. 7), yet they only deal with privacy to a limited extend or the GDPR and they leave most of the obligations to the users of these frameworks themselves. Yet there is no way for the user or data owner to control *WHAT* is being done with his private data.

To close this gap, we introduce an approach utilising homomorphic encryption. Private Data is encrypted from the very beginning or before being shared with unknown third parties utilising an encryption scheme with homomorphic properties. These properties enable any a third-party service provider to process this data without disclosing its clear text information. Data can be processed and sent back to the data owner for decryption if the latter explicitly gives his consent to the processing of this data in particular. Alternatively, data can be aggregated together with inputs from other parties. The resulting aggregation can be considered as anonymised and is hence no longer seen as private data under the rules of GDPR. Following protocols for decentralised decryption and Secure Multi-Party Computation, only these anonymised results are then available to service providers, who in return can continue their operation and service as usual. We present this as the ultimate goal in achieving privacy in the Internet of Things.

References

1. Amazon: Amazon Web Services Developer Guide. https://docs.aws.amazon.com/en_en/iot/latest/developerguide/iot-dg.pdf
2. Brakerski, Z., Gentry, C., & Vaikuntanathan, V. (2011). Fully homomorphic encryption without bootstrapping. Cryptology ePrint Archive, Report 2011/277. https://eprint.iacr.org/2011/277
3. Brakerski, Z., & Vaikuntanathan, V. (2011). Fully homomorphic encryption from ring-LWE and security for key dependent messages. In *Annual Cryptology Conference CRYPTO 2011: Advances in Cryptology - CRYPTO 2011* (pp. 505–524)
4. British Standards Institution. (2017). *EU General Data Protection Regulation 20 Steps to GDPR Compliance - A Methodical, Systematic and Logical Approach - A Whitepaper.* Tech. rep. https://www.bsigroup.com/LocalFiles/en-GB/CSIR/Resources/Whitepaper/UK-ENGB-CSIR-WP-20-steps-to-GDPR-PDF.pdf
5. Cavoukian, A. (2009). *Privacy by design: The 7 foundational principles.* Information and Privacy Commissioner of Ontario, Canada 5. https://www.iab.org/wp-content/IAB-uploads/2011/03/fred_carter.pdf
6. Damgå rd, I., Pastro, V., Smart, N., & Zakarias, S. (2011). Multiparty computation from somewhat homomorphic encryption. IACR Cryptology ePrint Archive 2011, 535. https://doi.org/10.1007/978-3-642-32009-5_38
7. Dijk van, M., Gentry, C., Halevi, S., & Vaikuntanathan, V. (2010). Fully homomorphic encryption over the integers. In *Annual International Conference on the Theory and Applications of Cryptographic Techniques EUROCRYPT 2010: Advances in Cryptology - EUROCRYPT 2010* (pp. 24–43)
8. Dix, A., Thüsing, G., Traut, J., Christensen, L., Etro, F., Aaronson, S. A., et al. (2013). EU data protection reform: Opportunities and concerns. *Intereconomics, 48*(5), 268–285. https://doi.org/10.1007/s10272-013-0470-y
9. Eclipse Foundation, Inc.: IoT developer survey 2019 results. (2019). https://iot.eclipse.org/resources/iot-developer-survey/iot-developer-survey-2019.pdf
10. European Commission: Regulation (EU) 2016/679 of the European parliament and of the council of 27 April 2016. *Official Journal of the European Union* L119/1 (2016). http://data.europa.eu/eli/reg/2016/679/2016-05-04
11. European Union. *Article 29 working party archives.* http://data.europa.eu/eli/reg/2016/679/2016-05-04
12. European Union. *European data protection board.* https://edpb.europa.eu/
13. European Union. (2012). Charter of fundamental rights of the European Union (2012/c 326/02)
14. Gentry, C. (2009). *A Fully Homomorphic Encryption Scheme.* Ph.D. thesis, Stanford University. https://crypto.stanford.edu/craig
15. Gentry, C., & Halevi, S. (2011). Implementing gentry's fully-homomorphic encryption scheme. In *Annual International Conference on the Theory and Applications of Cryptographic Techniques EUROCRYPT 2011: Advances in Cryptology - EUROCRYPT 2011* (pp. 129–148)
16. Guth, J., Breitenbücher, U., Falkenthal, M., Fremantle, P., Kopp, O., Leymann, F., et al. (2018). A detailed analysis of IoT platform architectures: Concepts, similarities, and differences. In *Internet of everything* (pp. 81–101). Singapore: Springer.
17. Kölsch, J., Heinz, C., Ratzke, A., & Grimm, C. (2019). Simulation-based performance validation of homomorphic encryption algorithms in the internet of things. *Future Internet, 11*(10). https://doi.org/10.3390/fi11100218. https://www.mdpi.com/1999-5903/11/10/218
18. Lee, H., & Kobsa, A. (2016). Understanding user privacy in internet of things environments. In *2016 IEEE 3rd World Forum on Internet of Things (WF-IoT)* (pp. 407–412). https://doi.org/10.1109/WF-IoT.2016.7845392
19. Microsoft: Microsoft Azure IoT Hub. https://docs.microsoft.com/de-de/azure/iot-hub/about-iot-hub
20. Paillier, P. (1999). Public-key cryptosystems based on composite degree residuosity classes. In J. Stern (Ed.), *Advances in Cryptology — EUROCRYPT '99* (pp. 223–238). Berlin: Springer.

21. Rivest, R. L., Shamir, A., & Adleman, L. (1978). A method for obtaining digital signatures and public-key cryptosystems. *Communications of the ACM, 21*(2), 120–126. http://doi.acm.org/10.1145/359340.359342
22. Sander, T., Young, A., & Yung, M. (1999). Non-interactive cryptocomputing for nc1. In *Proceedings of the 40th Annual Symposium on Foundations of Computer Science, FOCS '99* (p. 554). Washington: IEEE Computer Society.
23. Shafagh, H., Hithnawi, A., Burkhalter, L., Fischli, P., & Duquennoy, S. (2017). Secure sharing of partially homomorphic encrypted IoT data. In *Proceedings of the 15th ACM Conference on Embedded Network Sensor Systems, SenSys '17* (pp. 29:1–29:14). New York: ACM. http://doi.acm.org/10.1145/3131672.3131697
24. Tikkinen-Piri C., Rohunen A., & Markkula J. (2018) EU General Data Protection Regulation: Changes and implications for personal data collecting companies, Computer Law & Security Review, 34(1), pp:134–153, doi:https://doi.org/10.1016/j.clsr.2017.05.015. http://www.sciencedirect.com/science/article/pii/S0267364917301966
25. United Nations. (1948). Universal Declaration on Human Rights (UDHR). http://www.ohchr.org/EN/UDHR/Documents/UDHR_Translations/eng.pdf

Chapter 9
Mastering the IoT with the VICINITY Platform

Peter Kostelnik, Maria Koutli, and Viktor Oravec

1 Introduction to the VICINITY IoT Platform

The VICINITY is a platform that provides a set of tools and services that enable the integration and interaction of existing IoT infrastructures into a common framework for their interoperable operation. It offers a common language for IoT devices and services that aims to ease the development process of IoT applications, as well as the management of IoT objects, called Things, in a uniform way. This includes automatic registration, reading or setting properties values, execute actions, manage events, logging in and out of the peer to peer network, performing automatic discovery, and more.

In general, a Thing can be:

- a device: sensor, actuator, smart appliance, light bulb, weight scale, …
- a service: applications that use and process data from devices for monitoring, control, analysis, statistics, …

The architecture of the VICINITY platform is shown in Fig. 9.1, by using a use case application scenario with different connected IoT infrastructures and Value added services. As it can be seen the infrastructures are sending sensor measurements to Value Added Service 1 through the VICINITY Nodes and the VICINITY

P. Kostelnik
Intersoft A.S., Kosice, Slovakia
e-mail: peter.kostelnik@intersoft.sk

M. Koutli (✉)
Center for Research and Technology, Thessloniki, Greece
e-mail: mkoutli@iti.gr

V. Oravec
bAvenir, s.r.o., Bratislava, Slovakia
e-mail: viktor.oravec@bavenir.eu

© Springer Nature Switzerland AG 2021
C. Zivkovic et al. (eds.), *IoT Platforms, Use Cases, Privacy, and Business Models*,
https://doi.org/10.1007/978-3-030-45316-9_9

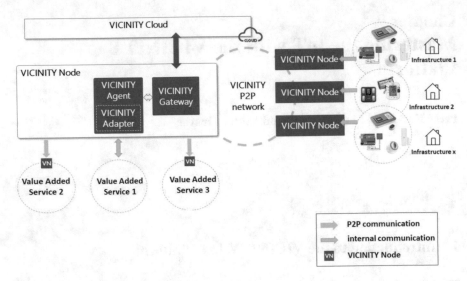

Fig. 9.1 The VICINITY platform architecture in a use case

Peer to Peer Network. Moreover, Value Added Service 1 is communicating with Value Added Services 2,3 for further data analysis, also through their VICINITY Nodes. The VICINITY Node of Value Added Service 1 is expanded in order to show the components it is composed of, namely the Adapter, the Agent and the Gateway. The VICINITY Cloud components are responsible for handling metadata information, as the description of Things and the access rules between them, while the Node components are responsible for the registration and the interactions between the Things.

The VICINITY Cloud is composed of three main components:

- **Neighbourhood Manager**: It provides all the functionality for User, Organisation registration, creation of Access Points, Friendships and Contracts and handling of visibility rights. It is composed of a dedicated API and a web interface.
- **Semantic Repository**: It stores all descriptions of registered Things in the VICINITY and their metadata information, such as name, owner and location.
- **Communication Server**: Its role is to setup communication channels between VICINITY Nodes and to control data exchange between them.

Each VICINITY Node contains the following components:

- **VICINITY Adapter** is the translator between IoT devices and services of a user/organisation and VICINITY common services. The Adapter is realised as REST service implementing prescribed API. The adapter implements a set of very basic services for interaction with the IoT Things that it exposes. However, it is not restricted to this set. It may implement its own API for interaction with its Things. This API is specified for each Thing in its formal description named *Thing description*.

Thing descriptions are used to create semantic models of Things to enable VICINITY interoperability services. They are also used by the component **VICINITY Agent** that performs translation from Adapter specific interaction API to the VICINITY common services.

- **VICINITY Agent** is the component that provides and automates the high-level tasks common for all Adapters to make adapter development lightweight and easy. The VICINITY Agent is able to manage multiple VICINITY Access Points and for each multiple Adapters, one Adapter per specific infrastructure. Common tasks supported by the Agent include:

  - providing uniform way of interaction with specific infrastructure
  - automatic discovery of things
  - automatic translation between infrastructure specific Thing identifiers and VICINITY identifiers
  - management of VICINITY credentials of Things for authentication into the VICINITY peer to peer network and automatic logging in/out of Things
  - automated event management
  - the Agent implements an API, that is used by Adapters as a bridge on the integration way of VICINITY nodes into the VICINITY peer to peer network

- **VICINITY Gateway API** implements all common VICINITY services and provides secure access and communication within the VICINITY peer to peer network

In order to exploit the features of the VICINITY platform, IoT devices and services need to become a part of the VICINITY ecosystem and interact. The following list lists the steps required for a user to integrate its own IoT infrastructures and services into the VICINITY platform:

- Registration of VICINITY User and Organisation, described in Sect. 2
- Registration of Access Points, described in Sect. 3
- Implementation of VICINITY Adapter, described in Sect. 4
- Creation of Thing Descriptions, described in Sect. 5
- Installation and configuration of the VICINITY Node, described in Sect. 6
- Automatic discovery and registration of Things, described in Sect. 7
- Creation of Contracts between Things, described in Sect. 8
- Interactions between Things, described in Sect. 9

Each section will provide a detailed description of the corresponding step followed by several hands-on examples.

2   Registration of VICINITY User and Organisation

The first step for the integration with the VICINITY is to create a user account in the Neighbourhood Manager. The simplest way to perform this is to use the VICINITY Neighbourhood Manager (NM) web interface. However, there is also an API that supports this step.

After visiting the web interface using the link https://vicinity.bavenir.eu/nm/, a user can select the "Register new company" option, which will open a registration form, as shown in Fig. 9.2.

In this form, the user should give information about the company, such as the name and the location, as well as personal information, such as the email and password of the account. After filling the required information in the form the user can proceed to registration.

The registration request will be received and validated by the platform administrator, who will inform the user by an email, stating whether the request was accepted or not. In case it is accepted (which is the most probable case), the user needs to validate his/her email by clicking the appropriate link, as shown in Fig. 9.3.

Fig. 9.2 The VICINITY registration form in Neighbourhood Manager

Dear representant of Example Org,

to activate your company account, please click on following link:

VALIDATION

This project has received funding from the European Union's Horizon 2020 Framework Programme for Research and Innovation under grant agreement no 688467.

Copyright © 2015 bAvenir, s.r.o . All rights reserved.

Fig. 9.3 Registration acceptance email

Fig. 9.4 Login page in Neighbourhood Manager

By clicking the "Validation" link the user will be transferred to the web interface, where a success message will be displayed and the user will be able to login to the platform by filling the username and password of the account, as shown in Fig. 9.4.

After logging in, the user can navigate to the different categories by using the sidebar menu. In order to be able to register devices or services in the VICINITY Neighbourhood Manager the user should have the appropriate roles. By clicking on the user's name or avatar, the Company's Profile is displayed, providing several options in a tab menu. When selecting the "Role Management" tab, the user can add or remove roles, by clicking the "Change Role" button as indicated with blue

Fig. 9.5 User role management in Neighbourhood Manager

circle in Fig. 9.5. In order to perform this action, the user should already have the administrator role, which is given by default to the user that registers the Organisation.

A user can have different roles in the VICINITY Neighbourhood Manager. Which action can be performed by which role is defined below:

- Device owner is able to perform actions related to devices.
- Service provider is able to perform actions related to services.
- Infrastructure operator is able to manage Contracts
- Administrator is the user that registers an organisation who gets this role by default. Administrator can modify the Organisation's properties, remove an Organisation and update Users roles.
- System integrator is able to create and manage Access Points.
- User is able to view the items of the Organisation and make very limited actions

Moreover, a user can change the account's privacy level by selecting the tab "User accounts", and then select his/her account. Next, in the User Profile the "Who can see?" option can be changed to any of the following:

1. Private Profile (default)
2. Visible for Friends or
3. Public Profile

Users can also set the privacy level of each registered Thing of their Organisation, according to their role, in order to allow the creation of Contracts between different Things. This will be further described in Sect. 8.

3   Registration of Access Points

Now, when the registration is successfully completed, the user can proceed to the second step of the integration process. This step requires registration of an Access Point for his/her IoT infrastructure. The Access Point will basically generate

Creating new Access Point

Access Point name

Enter name

Access Point type

VICINITY◉ Vicinity agent

SHAR-Q◯ Sharq adapter

Password

Enter password

Repeat Password

Repeat password

Submit   Cancel

Fig. 9.6 "Register access point" form

Name ⇕	AGID ⇕	Types ⇕	# Items ⇕		
aaa_example	882f5da7-4d43-4fb2-aba7-5006083ecb6d	generic.adapter.vicinity.eu VICINITY	0	✏	✖

Fig. 9.7 Access Points view

a unique id and store a password. This information will be then used for the VICINITY Agent's configuration. In this step, it is assumed that you have these two requirements fulfilled:

- Having registered an organisation in the VICINITY;
- Having a user with System Integrator role (this can be set in Neighbourhood Manager as described in Sect. 2).

In the following you find the steps as guidelines for registering an Access Point:

- First, log in to the VICINITY platform;
- Then navigate to the Access Points option, located at the left sidebar menu;
- In the Access Points section click on "Create Access Point", located at the top-right corner; the form shown in Fig. 9.6 will pop up.
- Fill in the form following these steps:

  - Type a new name for identifying the Access Point,
  - Select the type of Access Point, for VICINITY projects select VICINITY Agent,
  - Type a password, it will be important to remember the password,
  - Click on Submit button

After successful registration of the Access Point, a new item will appear in the Access Points view, as shown in Fig. 9.7. The new access point has a name, an AGID, which is the unique id that was assigned to it by the Neighbourhood Manager, a type, which is VICINITY and the items that are registered under this access point, which are currently zero. The AGID and the password are the necessary credentials to configure the VICINTY Agent in the infrastructure.

4   Implementation of VICINITY Adapter

The VICINITY Adapter is a software component that is used to expose Things to the VICINITY platform. It is basically an interface between the IoT infrastructure (software or hardware) and the VICINITY world. The adapter should be implemented in such a way, so that it translates the infrastructure's vocabulary to the syntactic and semantic syntax that is needed for operations within the VICINITY. More specifically, the Adapter has two main functionalities:

- Provide the interaction patterns for a Thing
- Provide the Thing Description

The Adapter can be implemented as a RESTful web service, which exposes three types of resources, namely, Properties, Actions and Events. For each type, a certain set of interactions is allowed to be implemented by the Adapter, which can be described with the following endpoints:

- Properties

    - Get Property—GET /objects/oid/properties/pid
    - Set Property—PUT /objects/oid/properties/pid

- Actions

    - Start an Action—POST /objects/oid/actions/aid
    - Cancel an Action—DELETE /objects/oid/actions/aid/tasks/tid

- Events

    - Receive Events—PUT /objects/oid/events/eid

The above endpoints will be described in detail in the Thing Description, that is covered in the following section. The Adapter is responsible to provide this description to the Agent in order to allow the Things' registration, update or removement. This can be done in two ways, according to the Agent's configuration. One option is that the Adapter stores the Thing Description and sends it to the Agent for the first time and each time when the Thing Description is changed. In the second option the Adapter has to implement one more endpoint:

- Get Objects—GET/objects

In this case, the adapter should return the Thing Description as a response to the call of the above endpoint. This call is made by the Agent, every time it is started, in order to perform the appropriate registration, update or deletion operations for the Things that are under the Adapter.

5 Creation of Thing Descriptions

The main responsibility of the VICINITY Adapter is to describe all Things it exposes to the VICINITY platform. The Thing Description is a JSON document, that enables the VICINITY to understand what Things are exposed and what are their capabilities. The Thing Description is automatically translated into a semantic model managed by the VICINITY semantic platform described in Chap. 5. The semantic model of things enables the VICINITY platform to automatically understand what is the thing and how to interact with it. It can also be used by rich semantic interoperability services provided by the VICINITY. To be able to translate Thing Description into the semantic model, it must contain various semantic annotations that will allow automatic linking to the VICINITY adapters ontology,[1] described briefly in Chap. 5.

5.1 Thing Description Overview

Things are described using the top-level JSON schema of Thing Description. The top-level schema can be divided into three parts:

- *IDs and metadata*: Thing identifier—oid (stands for object identifier), name, type, version, keywords and located-in.
- *Interaction patterns*: Resources, that are exposed by Thing. The available interaction patterns are: properties, actions and events. Properties are used to read or write a value into the Thing (for example, get/set value on weight scale). Actions are used to start a process that may take some time to finish (for example, fade off the light in the room). Events are used to open or subscribe a channel, where data may be pushed and received (continually publish or receive sensor measurement values).
- *Services requirements*: Services must specify which are interactions patterns that they can work with. For example, it would not be interesting to monitor a door sensor with a service that collects the energy production of photovoltaics (PVs).

---

[1] http://iot.linkeddata.es/def/adapters.

JSON Schema of Thing Description

```
{
  "oid": {type: String, mandatory: true},
  "name": {type: String, mandatory: true},
  "type": {type: String, mandatory: true},
  "version": {type: String, mandatory: false},
  "keywords": {type: Array[String], mandatory: false},
  "located-in": {type: Array[Object], mandatory: false},
  "actions" : {type: Array[Object], mandatory: true},
  "events": {type: Array[Object], mandatory: true},
  "properties": {type: Array[Object], mandatory: true},
  "requirements": {type: Object, mandatory: false}
}
```

In the following we provide the step-by-step guide how to describe a thing following the schema introduced above. A simple light bulb is used as a demonstration example.

5.2 Minimal Configuration of a Thing

The minimal configuration requires very basic information about a thing. This information includes oid, name, type of the thing and interaction patterns. Following the schema, for the light bulb example the minimal Thing Description could look like:

Minimal Thing Description

```
{
  "oid": "lightbulb-1",
  "name": "MyFirstLightBulb",
  "type": "adapters:Lightbulb",
  "actions" : [],
  "events": [],
  "properties": []
}
```

where

- **oid** is a unique identifier of thing known by Adapter of local infrastructure. There are no specific restrictions to this field; it is up to Adapter developer to create and manage it. In this example Adapter must internally know the Light bulb under its internal, infrastructure specific local oid "lightbulb-1". When the Thing is discovered and exposed to the VICINITY, it receives another unique identifier (**VICINITY oid**) that is used for all communications in the VICINITY neighbourhood. Adapter of local infrastructure always uses the **oid** when describing or interacting with its own Things. **VICINITY oid** is used to interact with remote things exposed by Adapters of third-party infrastructures with which the local infrastructure interacts.

- **name** is a human readable name of thing, that will be displayed in the VICINITY Neighbourhood Manager.
- **type** is annotation to the VICINITY Adapters ontology,[2] which specifies the semantic type of the Thing. There are two generic root types **core:Device** and **core:Service**. Semantic annotation is always in the form **prefix:value**, where prefix points to specific ontology and value is specific ontology class or instance. In our example, we use annotation to the class **Lightbulb** defined in the Adapters ontology: **adapters:Lightbulb**. To describe the service, it is required to use the class annotation **core:Service**.
- **Interaction patterns (properties, actions and events)**. For validation purposes, they must be included in Thing Description, even though they are empty.

5.3   Location Metadata Information

Location adds geographical or environmental metadata that can be understood by the VICINITY semantic interoperability services. Location information may contain as much metadata as needed. One example providing information about country and city, where Thing is located, is given below:

Specifying Thing Location in Thing Description

```
{
  "located-in ": [
    {
      "location_type": "s4city:City",
      "label": "Bratislava",
      "location_id": "http: //dbpedia.org/resource/Bratislava"
    },
    {
      "location_type": "s4city:Country",
      "label": "Slovakia",
      "location_id": "http: //dbpedia.org/resource/Slovakia"
    }
  ]
}
```

---

[2]http://iot.linkeddata.es/def/adapters/index-en.html#desc.

Each location metadata object contains the following information:

- **location_type**: Semantic annotation to the class **City** in the location ontology SAREF4CITY.[3] Currently, there are four available classes defined in this ontology: **Country, City, District** and **Neighbourhood**. They are all subclasses of the class **AdministrativeArea**. This field is mandatory.
- **label**: Human readable name of location. This is also a mandatory field.
- **location_id**: Location identifier, which refers to valid URI representing the RDF identifier of resource. It is used by semantic interoperability services to identify the same location metadata in different Thing Descriptions and hence, discover the things subject to this location. This is not a mandatory field.

5.4   Thing's Interaction Patterns

By adding interaction patterns in the Thing Description, the VICINITY platform understands how to automatically interact with the Thing. As described in Sect. 5.1 there are three available interaction patterns: properties, actions and events. Each interaction pattern in Thing description is specified with its own identifier.

Interaction Patterns in Thing Description for Light Bulb Example

```
"properties" : [
   "pid": "is-on",
   ......
],
"actions": [
"aid": "fade-in",
....
],
"events": [
"eid": "power-consumption"
]
```

The property with the identifier "is-on" will tell if the light bulb is switched on or off. The action with the identifier "fade-in" dims the light of the light bulb in. The event "power-consumption" will periodically send notifications about actual average of power consumption of the light bulb. The rest of the descriptions for each interaction pattern will be explained in the following.

---

[3]https://w3id.org/def/saref4city.

5.4.1   Semantic Annotations

Each interaction pattern is annotated to the VICINITY adapters ontology, that enables a common representation and understanding what property or quality of the world the pattern monitors or affects. Properties and events may monitor some world property, actions may affect some world property. In the Light Bulb example the property "is-on" monitors property of the world representing if thing is on or off. Thus it is annotated to adapters:OnOff property in the ontology. The action "fade-in" affects the dimming level of the bulb, and hence, it is annotated to adapters:DimmingLevel property in the ontology. VICINITY Adapters ontology describes the hierarchy of possible world properties, that can be monitored by Thing's properties and events and affected by Thing's actions.

5.4.2   Data Schema

The interaction patterns exchange data. When the interaction pattern is specified for reading, it returns output data (e.g. property value). When the interaction pattern is specified for writing (e.g. *set* property or *execute* action), it consumes input data, telling, to what value should be property set or what are the input parameters required to execute action. As a result, data is returned as notification on property set or action execution status.

There are no restrictions on structure of input/output data. Consequence is that input/output data may be very specific and different for each interaction pattern. It is required to describe them in order to understand, what data interaction pattern consumes and what response is expected. For this purpose Data Schema is introduced. It specifies how to describe expected input and output of pattern.

VICINITY Data Schema is based on the W3C Thing Description typed system.[4] The root of input/output data schema is always a JSON object, which may embed another JSON objects or arrays. So, its type is always defined as "object".

Data Schema for Describing Pattern Input and Output

```
{
  "type": "object",
  "field": [
    {
      "name": "person-name",
      "predicate": "core:value",
      "schema": {
        "type": "string"
      }
    }
```

---

[4]https://www.w3.org/TR/wot-thing-description/#type-system-section.

```
    },
    {
      "name": "address",
      "schema": {
        "type": "object",
        "field": [
          {
            "name": "street",
            "schema": {
              "type": "string"
            }
          },
          {
            "name": "country",
            "schema": {
              "type": "string"
            }
          }
        ]
      }
    }
  ]
}
```

Expected JSON payload implementing this data schema:

JSON Payload

```
{
  "person-name": "some person name",
  "address": {
    "street": "some street",
    "country": "some country",
  }
}
```

The predicate "core:value" in Data Schema represents the semantic annotation of the real value of the JSON object. This annotation is important for the VICINITY semantic interoperability services to understand which field of the JSON object represents the real value of this object. For example, if this JSON object is returned as response on the request to get person name, the field "person-name" represents the real value for this property.

5.4.3   Properties

Each property is described with the following field:

- **pid**: Identifier of property. It must be unique within Thing Description.
- **monitors**: Semantic annotation of monitored world property/quality in the VICINITY adapters ontology.
- **read_link**: This field is composed of two other fields "output" and "href". The field "output" contains the description of expected output returned by Adapter for this property, defined by Data Schema. The endpoint "href" is the endpoint of the REST service GET that must be implemented by Adapter in order to return value of this property. This endpoint returns property value in the format described in the field "output".
- **write_link**: This field is composed of three other fields "input", "output" and "href". The field "input" contains description of expected input required for the property value to be set and the "output" is the expected output returned by Adapter when the property value is set successfully (defined by Data Schema). The endpoint "href" specifies the href endpoint of the REST service PUT that must be implemented in Adapter in order to return value of this property. This endpoint consumes JSON described in the field "input" and returns response in the format described in the field "output".

In our Light Bulb example the property "is-on" described in Sect. 5.4 is readable and writable property. By reading it, we will receive Boolean information if the light bulb is on or off and its power consumption in the case it is on. By writing Boolean value to it, we will be able to switch the light bulb on or off.

Description of Light Bulb Property "is-on"

```
{
  "properties": [
    {
      "pid": "is-on",
      "monitors": "adapters:OnOff",
      "read_link": {
        "output": {
          "type": "object",
          "field": [
            {
              "schema": {
                "type": "boolean"
              },
              "predicate": "core:value",
              "name": "is-on"
            },
            {
              "schema": {
                "type": "double"
```

```
            },
            "name": "consumption"
          }
        ]
      },
      "href": "/objects/lightbulb-1/properties/is-on"
    },
    "write_link": {
      "input": {
        "type": "object",
        "field": [
          {
            "schema": {
              "type": "boolean"
            },
            "name": "is-on"
          }
        ]
      },
      "output": {
        "type": "object",
        "field": [
          {
            "schema": {
              "type": "boolean"
            },
            "predicate": "core:value",
            "name": "is-on"
          }
        ]
      },
      "href": "/objects/lightbulb-1/properties/is-on"
    }
  }
]
}
```

Note, that "href" in read/write_link provides the specification of Adapter API endpoints, that must be implemented by Adapter. Agent will use these endpoints to interact with Adapter. As a response to reading this property, Adapter should respond with payload whose expected format is defined in the field "output". For this example it is:

Adapter Response to Reading the Light Bulb Property "is-on"

```
{"is-on": true, "consumption": 34.2}
```

Adapter Response to Setting Value of the Property "is-on"

```
{"is-on": false}
```

5.4.4 Actions

Each action can be described with the following fields:

- *aid*: Identifier of action. Must be unique within Thing Description.
- *affects*: Semantic annotation that points to the affected property from the VICINITY adapters ontology.
- *write_link*: This field contains three fields: "input", "output" and "href". The field "input" contains the description of expected input to execute action and "output" is the output returned by Adapter as a result of executing the action (both defined by Data Schema). The endpoint "href" specifies the href endpoint of REST service POST that must be implemented in Adapter in order for this action to be executed. Actions are used to start a process that may take some time to finish (e.g. fading in the Light bulb will take some time to perform). Life-cycle of action management is a bit more complex and will be explained in Sect. 9.
- *read_link*: This field contains the fields "output" and "href". The "output" contains description of expected output returned by Adapter for actual status or value of this action (defined by Data Schema). The endpoint "href" is the href endpoint of REST service GET that must be implemented in Adapter in order to return status of this action. The endpoint returns action value in the format described in "output".

As demonstrated in Sect. 5.4 we added the action "fade-in", to our light bulb. For the execution of this action the expected input is the light intensity with a value of type "double".

As the output of successful action execution, the execution status is returned. By reading the action value, the response will contain the current value of the light intensity of the light bulb of type "double".

Description of Light Bulb Action "fade-in"

```
{
  "actions": [
    {
      "aid": "fade-in",
      "affects": "adapters:DimmingLevel",
      "write_link": {
```

```
        "input": {
          "type": "object",
          "field": [
            {
              "schema": {
                "type": "double"
              },
              "name": "light-intensity"
            }
          ]
        },
        "output": {
          "type": "object",
          "field": [
            {
              "schema": {
                "type": "string"
              },
              "name": "status"
            }
          ]
        },
        "href": "/objects/lightbulb-1/actions/fade-in"
      },
      "read_link": {
        "output": {
          "type": "object",
          "field": [
            {
              "schema": {
                "type": "double"
              },
              "predicate": "core:value",
              "name": "light-intensity"
            }
          ]
        },
        "href": "/objects/lightbulb-1/actions/fade-in"
      }
    }
  }
]
}
```

Example of Expected Input for the Action "fade-in"

```
{"light-intensity": 45.6}
```

Example of Expected Response Reporting the Status of the Action "fade-in"

```
{"status": "running"}
```

Example of Expected Response to Reading the Value of the Action "fade-in"

```
{"light-intensity": 34.2}
```

5.4.5   Events

Events use slightly different description in comparison with properties and actions. They do not contain read or write links. In the VICINITY event mechanism, Adapter is responsible to periodically push the events into event channel responsible for this event. Each event is described with the following fields:

- *eid*: Identifier of event. Must be unique within Thing Description.
- *monitors*: Semantic annotation of monitored world property/quality in the VICINITY adapters ontology.
- *output*: Expected schema of payload for this event.

We will add the event "power-consumption", to our light bulb. The light bulb will periodically send events containing its actual power consumption.

Description of *Light Bulb* Event "power-consumption"

```
{
  "events": [
    {
      "eid": "power-consumption",
      "monitors": "adapters:MeanPowerConsumption",
      "output": {
        "type": "object",
```

```
        "field": [
            {
                "schema": {
                    "type": "double"
                },
                "name": "power-consumption"
            }
        ]
    }
  }
 ]
}
```

Following the payload "output" data schema, the expected format of the published events is as follows:

Example of Expected Output of the Event "power-consumption"

```
{"power-consumption": 34.2}
```

5.4.6   Putting It All Together…

Now, when we have all information, we are in the position to create the final Thing Description of the light bulb.

The Final Thing Description of Light Bulb

```
{
  "oid": "lightbulb-1",
  "name": "MyFirstLightBulb",
  "type": "adapters:Lightbulb",
  "located-in ": [
    {
      "location_type": "s4city:City",
      "label": "Bratislava",
      "location_id": "http: //dbpedia.org/resource/Bratislava"
    },
    {
      "location_type": "s4city:Country",
      "label": "Slovakia",
      "location_id": "http: //dbpedia.org/resource/Slovakia"
    }
  ],
```

```
"actions": [
  {
    "aid": "fade-in",
    "affects": "adapters:DimmingLevel",
    "write_link": {
      "input": {
        "type": "object",
        "field": [
          {
            "schema": {
              "type": "double"
            },
            "name": "light-intensity"
          }
        ]
      },
      "output": {
        "type": "object",
        "field": [
          {
            "schema": {
              "type": "string"
            },
            "name": "status"
          }
        ]
      },
      "href": "/objects/lightbulb-1/actions/fade-in"
    },
    "read_link": {
      "output": {
        "type": "object",
        "field": [
          {
            "schema": {
              "type": "double"
            },
            "predicate": "core:value",
            "name": "light-intensity"
          }
        ]
      },
      "href": "/objects/lightbulb-1/actions/fade-in"
    }
  }
],
"events": [
  {
    "eid": "power-consumption",
    "monitors": "adapters:MeanPowerConsumption",
    "output": {
      "type": "object",
      "field": [
        {
```

```
              "schema": {
                "type": "double"
              },
              "name": "power-consumption"
          }
        ]
      }
    }
  ],
  "properties": [
    {
      "pid": "is-on",
      "monitors": "adapters:OnOff",
      "read_link": {
        "output": {
          "type": "object",
          "field": [
            {
              "schema": {
                "type": "boolean"
              },
              "predicate": "core:value",
              "name": "is-on"
            },
            {
              "schema": {
                "type": "double"
              },
              "name": "consumption"
            }
          ]
        },
        "href": "/objects/lightbulb-1/properties/is-on"
      },
      "write_link": {
        "input": {
          "type": "object",
          "field": [
            {
              "schema": {
                "type": "boolean"
              },
              "name": "is-on"
            }
          ]
        },
        "output": {
          "type": "object",
          "field": [
            {
              "schema": {
                "type": "boolean"
              },
              "predicate": "core:value",
```

```
                "name": "is-on"
            }
        ]
    },
    "href": "/objects/lightbulb-1/properties/is-on"
    }
  }
 ]
}
```

6   Installation and Configuration of a VICINITY Node

This section should serve as an installation guide for a VICINITY node. Following the steps below the following components will be installed:

- VICINITY Gateway API
- VICINITY Agent
- VICNITY Adapter with two Things

6.1   Installation of VICINITY Gateway API

We will start with the VICINITY Gateway API installation. For this process the following steps are required:

Clone the GitHub Repository

```
$ cd /path/to/the/directory
$ git clone git@github.com:vicinityh2020/vicinity-gateway-api.git
```

Compile the VICINITY Gateway API Using Maven

```
$ cd vicinity-gateway-api
$ mvn clean package
```

This step is optional. The latest stable version of the VICINITY Gateway API is included as a binary in the GitHub repository, but it is possible to compile your own from the source with maven.

Create Dedicated System User for the VICINITY Gateway API

```
$ adduser --system --group --home /opt/ogwapi --shell /bin/sh
    ↪ ogwapi
```

Place Everything in /opt/ogwapi

```
$ mkdir /opt/ogwapi
$ mkdir /opt/ogwapi/log
$ mkdir /opt/ogwapi/config
$ cp -r ./config /opt/ogwapi
$ cp target/ogwapi-jar-with-dependencies.jar /opt/ogwapi/gateway.
    ↪ jar
$ chown -R ogwapi:ogwapi /opt/ogwapi
$ chmod u+x /opt/ogwapi/gateway.jar
```

Run the VICINITY Gateway API

```
$ cd /opt/ogwapi
$ su - ogwapi -c "nohup java -jar gateway.jar &"
```

It is also possible to run the gateway as a service. For more information see the end of this section.

Check the Running API; Expected Content of nohup.out File

```
CONFIG: HTTP Basic challenge authentication scheme configured.
Sep 08, 2018 2:19:45 PM org.restlet.engine.connector.
    ↪ NetServerHelper start
INFO: Starting the internal [HTTP/1.1] server on port 8181
Sep 08, 2018 2:19:45 PM org.restlet.Application start
INFO: Starting eu.bavenir.ogwapi.restapi.Api application
Sep 08, 2018 2:19:45 PM org.restlet.engine.application.
    ↪ ApplicationHelper start
```

```
FINE: By default, an application should be attached to a parent
  ↪ component in order to let application's outbound root
  ↪ handle calls properly.
Sep 08, 2018 2:19:45 PM eu.bavenir.ogwapi.restapi.RestletThread
  ↪ run
FINE: HTTP server component started.
```

As mentioned above, the VICINITY Gateway API can be started as a service. In Linux machines these are the steps required:

Create File in /etc/systemd/system with the Following Content

```
[Unit]
Description = ogwapi service
After = network.target

[Service]
Type = forking
ExecStart = /usr/local/bin/ogwapi.sh start
ExecStop = /usr/local/bin/ogwapi.sh stop
ExecReload = /usr/local/bin/ogwapi.sh reload
SuccessExitStatus=143
Restart=always

[Install]
WantedBy=multi-user.target
```

For this purpose you can use either **nano** or **vim** text editor and safe the file under name *ogwapi.service*. For example, with **vim** text editor the command would be:

```
vim /etc/systemd/system/ogwapi.service.
```

Create a File ogwapi.sh in /usr/local/bin/ with the Following Content

```
#!/bin/sh
SERVICE_NAME=ogwapi
PATH_TO_JAR_FOLDER=/home/andrej/ogwapi
PATH_TO_JAR=$PATH_TO_JAR_FOLDER/ogwapi.jar
PID_PATH_NAME=/tmp/ogwapi-pid
cd $PATH_TO_JAR_FOLDER
case $1 in
  start)
    echo "Starting_$SERVICE_NAME_..."
```

```
  if [ ! -f $PID_PATH_NAME ]; then
    nohup java -jar $PATH_TO_JAR >> $PATH_TO_JAR_FILE_FOLDER/
       ↪ ogwapiService.out 2>&1&
    echo $! > $PID_PATH_NAME
    echo "$SERVICE_NAME started ..."
  else
    echo "$SERVICE_NAME is already running ..."
  fi
  ;;
  stop)
   if [ -f $PID_PATH_NAME ]; then
    PID=$(cat $PID_PATH_NAME);
    echo "$SERVICE_NAME stoping ..."
    kill $PID;
    echo "$SERVICE_NAME stopped ..."
    rm $PID_PATH_NAME
   else
    echo "$SERVICE_NAME is not running ..."
   fi
   ;;
  restart)
  if [ -f $PID_PATH_NAME ]; then
    PID=$(cat $PID_PATH_NAME);
    echo "$SERVICE_NAME stopping ...";
    kill $PID;
    echo "$SERVICE_NAME stopped ...";
    rm $PID_PATH_NAME
    echo "$SERVICE_NAME starting ..."
    nohup java -jar $PATH_TO_JAR >> $PATH_TO_JAR_FILE_FOLDER/
       ↪ ogwapiService.out 2>&1&
    echo $! > $PID_PATH_NAME
    echo "$SERVICE_NAME started ..."
  else
    echo "$SERVICE_NAME is not running ..."
  fi
  ;;
esac
```

The command for creating this file with the **vim** text editor is as follows:

```
sudo vim /usr/local/bin/ogwapi.sh.
```

6.2   Installation of VICINITY Agent and Adapter

We provide a very simple Adapter example as a playground for first run and testing. The Adapter example is part of the VICINITY Agent installation.

- Download the prepared Adapter example from the VICINITY Agent GitHub repository[5];
- In *releases* tab of the GitHub repository, find the last release and download attached file adapter-build-x.y.z.zip, where x.y.z is the version of the actual release;
- Unzip it.

Examples Adapter exposes two Things: example-thing-1 and example-thing-2. You can find their thing descriptions in the file:

```
adapter-build-x.y.z/objects/example-objects.json
```

6.3 Agent Configuration

The VICINITY Agent is designed to manage multiple different VICINITY *Access Points* (described in Sect. 3). For each *Access Point*, it is possible to manage multiple infrastructures. For each specific infrastructure, a separate Adapter must be implemented. The VICINITY Agent automates common tasks with respect to this structure of access points and adapters. In order to inform Agent about this structure, it must be properly configured.

Agent configuration contains the set of JSON configuration files, one per *Access Point*. For each *Access Point* it is required to:

- set the *Access Point* credentials obtained in *Access Point* registration
- configure adapters used by each *Access Point*
- for each adapter,

  - give unique Adapter identifier
  - configure endpoint, where Adapter REST service API is available
  - configure events, including automatic opening of event channels for local Things exposed by Adapter and automatic subscription to event channels provided by remote Things (belonging to another Access Point)

Example of *Access Point* Configuration in the Agent

```
{
  "credentials": {
    "agent-id": "access point agid",
    "password": "access point password"
  },
  "adapters": [
```

```
{
    "adapter-id": "unique id of adapter",
    "endpoint": "adapter REST service endpoint",
    "events": {
      "channels": [
          {
            "infrastructure-id": "internal identifier of
                ↪ publisher",
            "eid": "object event identifier"
          }
      ],
      "subscriptions": [
          {
            "infrastructure-id": "internal identifier receiver",
            "oid": "VICINITY id of remote publisher",
            "eid": "object event identifier"
          }
      ]
    }
  }
]
}
```

7   Automatic Discovery and Registration of Things

The VICINITY Agent is able to handle multiple different *Access Points*.

The Agent automates complex tasks, such as mapping between infrastructure specific local Thing identifiers and common VICINITY identifiers, automatic authentication into the VICINITY network, event channel management and the automatic Thing discovery process.

In the automatic discovery process, Things exposed by the Adapter are automatically processed and registered in the VICINITY neighbourhood. This includes creation of VICINITY oid, that will be used to interact with the Thing on the VICINITY platform level, creation of credentials to authenticate *Things* in the VICINITY network and also rich semantic model in the VICINITY semantic platform.

Note, that we need to deal with two Thing identifiers:

- **Local Thing identifier** *local oid*: local, infrastructure specific identifier, managed by Adapter. When Adapter is triggering any interaction within the VICINITY, it always uses these internal identifiers for Things it exposes.
- **Common VICINITY identifiers** *VICINITY oid*: identifiers that are created in the automatic discovery process. In the VICINITY, each Thing has unique VICINITY oid, under which it is known in the whole VICINITY platform.

Mapping between local and VICINITY oid is automatically managed by the Agent.

The automatic Thing discovery manages all changes made in the list of Things exposed by Adapter: creation of new Things, changes and updates of already existing Things and removal of missing Things (that are not exposed anymore). Every CRUD (Create, Read, Update and Delete) operation on Thing exposed by the Adapter in the discovery process affects the internal Agent configuration, the VICINITY Neighbourhood and the VICINITY semantic models.

However, the automatic discovery is automated in the Agent. From Adapter point of view, the discovery is very easy. Literally it is enough to expose Things behind the Adapter to the Agent. The full complex process is performed automatically.

The Things discovery process is always performed per one concrete Adapter. The Adapter is responsible to present the actually exposed list of its Thing Descriptions in the form:

JSON Format for Exposed List of Things

```
{
 "adapter-id": "unique adapter identifier",
 "thing-descriptions": Array[Thing Description Object]
}
```

- **adapter-id**: Unique adapter ID of specific adapter specified in *Access Point* configuration. As the Agent may serve multiple Adapters, each Adapter is responsible for providing its own Adapter identifiers, so the Agent knows, which Adapter is to be discovered.
- **thing-descriptions**: Array of JSON objects that represent Thing descriptions and that are described in Sect. 5.

There are two scenarios for automatic Things discovery: passive and active.

7.1   Passive Discovery

The Agent fetches the Thing Descriptions from the Adapter when it starts or when the Adapter rediscovery is requested. This is default scenario.

In this case, the Adapter must implement the REST endpoint GET: /objects. This endpoint must respond with the actual list of Thing Descriptions in the form described above.

7.2   Active Discovery

Adapter pro-actively sends the Thing Descriptions into the Agent. In this case, the Agent does not fetch any discovery information from the Adapter. It is Adapter's responsibility to provide it.

Active discovery process can be activated in the *Access Point* configuration JSON file setting "active discovery" to true.

Access Point Configuration for Active Discovery

```
{
  "credentials": {},
  "adapters": [
    {
      "adapter-id": "unique id of adapter",
      "endpoint": "adapter REST service endpoint",
      "active-discovery": true
    }
  ]
}
```

The Adapter must push the actual list of Thing Descriptions into the Agent, using Agent API service PUT: /objects. The body of PUT request must be Thing Descriptions JSON file in the form described in Sect. 5. Active discovery can be used on demand. Thing Descriptions may be updated on the fly by Adapter, in the case the list of its Things changes.

8   Creation of Contracts Between Things

Things that need to interact must be successfully discovered and obtain their common VICINITY oid. To allow interaction between two Things, it is required that they can see each other. In other words, either they belong to the same organisation or there is a contract between them.

Currently there is only one contract type "Service Request". In this type of contract, services and devices from one organisation can request to use a service from another organisation. This service monitors those items that are added to the contract by the first organisation.

In order to create a contract, these steps must be followed using the VICINITY Neighbourhood Manager user interface:

- A user with the role *Infrastructure Operator* clicks the button *Request service* from the desired service in the services view. A new window with *Contract settings* opens where the *Infrastructure Operator* needs to:
  - Select Things from a list with the fitting devices and services of his organisation; Things can belong to different users.
  - Approve the terms and conditions.
- A notification is sent to the *Service Provider*.
- Once the *Service Provider* approves the contract, the service can start sharing data with the items belonging to the *Infrastructure Operator*, if there are any. All other items that belong to different users will not be part of the contract until their owners give consent. Therefore, the next step is to wait until all users approve or reject the contract. All of them will be notified to do so.

In the end the contract is created between the *Infrastructure Operator*, the *Service Provider* and all other users that gave consent to share data. Any time a user can cancel the contract.

9   Interactions Between Things

The high-level concept of interactions in the VICINITY platform is, that each interaction happens always between two Things. That means:

- If Thing(1) needs to interact with Thing(2), in the VICINITY neighbourhood, the proper permission for this interaction must be set. Concretely, a contract between Thing(1) and Thing(2) must be created.
- Each interaction must be authenticated with credentials of Thing(1), that triggers the interaction. Credentials of the Thing are obtained in the automatic discovery process and are automatically managed by the Agent.

Conceptually, the VICINITY platform enables interaction between two Things. However, it is up to Adapter developer, how these interactions are implemented. In order to perform the interaction, prescribed sequence of calling Agent API services must be complied. Thing interaction patterns (properties, actions, events) have different nature. For this reason, for each interaction pattern, the interaction use-cases are different. All typical interaction use-cases will be described in this section on the conceptual level (Fig. 9.8).

From the Adapter's point of view, the interaction is triggered by **local Thing(1)** managed by its Adapter(1). Thing(1) needs to interact with **remote Thing(2)** exposed by different Adapter(2) on the different VICINITY *Access point*. Note, that Adapter always uses local oid when managing its own Things. Target remote Thing is always addressed using common **VICINITY oid** known for the whole platform. High-level scenario of interaction between local Thing(1) and remote Thing(2) can be described from the Adapter point of view as follows:

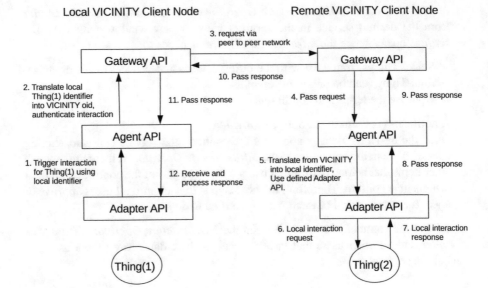

Fig. 9.8 The VICINITY Node-to-Node communication

1. Adapter(1) triggers interaction for its local Thing(1) by calling corresponding Agent API service.
2. For each interaction, Adapter provides

   - its own **adapter-id**
   - **local oid of Thing(1)**, which triggers the interaction
   - common **VICINITY oid of remote Thing(2)**, with which **Thing(1)** needs to interact

3. The Agent receives the attempt for interaction and finds common VICINITY representation of Thing(1) and its credentials. The Agent automatically translates this interaction into the VICINITY common service, authenticates it and executes it via the local Gateway API available in the VICINITY Node, where the Adapter runs.
4. The Gateway API performs interaction via the VICINITY peer to peer network.
5. Request arrives into the remote VICINITY node, where it reaches its Gateway API.
6. The Gateway API on the remote node forwards this request to the Agent of the node.
7. The Agent automatically translates interaction request into local oid of target Thing(2) and uses corresponding API of Adapter(2) to interact with Thing(2).
8. Adapter(2) responds to request and response is returned back to the local Adapter(1) containing Thing(1).

9.1 *Properties*

Use-Case 1: Request of Thing(1) in the Local Adapter(1) to Read Property of Remote Thing(2) in Adapter(2)

1. Adapter(1) calls the Agent API service GET /remote/objects/vicinity-oid/properties/pid, where

   - vicinity-oid is the VICINITY identifier of remote Thing(2)
   - pid is the identifier of Thing(2) property, specified in its Thing Description
   - Adapter(1) identifier and local oid of Thing(1) are attached in request headers

2. The request reaches the Gateway API and the Agent on the remote VICINITY node
3. The Agent uses the Thing Description of Thing(2) to call specified corresponding Adapter(2) endpoint to read property
4. The response provided by Adapter(2) is returned back to Adapter(1)

Use-Case 2: Request of Thing(1) in Adapter(1) to Set Value of a Property of Remote Thing(2) in Adapter(2)

1. Thing(1) calls the Agent API service PUT /remote/objects/vicinity-oid/properties/pid, where

   - vicinity-oid is the VICINITY identifier of target remote Thing(2) and
   - pid is identifier of its property, provided in its Thing Description
   - Adapter(1) identifier and local identifier of Thing(1) are attached in request headers

2. The request reaches the Gateway API and the Agent on the remote VICINITY node
3. The Agent uses the Thing Description of Thing(2) to call specified corresponding Adapter(2) endpoint to set property
4. The response provided by Adapter(2) is returned back to Adapter(1)

9.2 Actions

In the VICINITY, actions are used to start a process that may take some time to finish. For example, the action "fade-in" in the Light Bulb example, may take couple of seconds to be completed. More complex actions may take minutes or even hours. Taking this into account, the high-level life-cycle of action is little bit more complex and can be described as follows.

Steps for Executing Actions

- Thing(1) on Adapter(1) executes an action of remote Thing(2) on Adapter(2)
- Remote Gateway API, to which Adapter(2) belongs, generates the specific, unique identifier of the task *task-id*, which will be executed by target Thing(2) on Adapter(2). The task represents the single execution of action. The Gateway API holds the queue of tasks being actually executed on the VICINITY node. Each specific task may run only once in time. If there are multiple requests for execution of same action, they are executed one by one as they come into the queue. Once the task is finished, the next one in the queue is executed.
- Action is now being executed on Thing(2) and *task-id* is returned back to Adapter(1), which requested this execution.
- Adapter(2) continually updates the status of action execution. Actual status and action execution output is kept on the remote Gateway API.
- Adapter(1) continually reads the status of task using corresponding *task-id* generated for the request.

Use-Case 1: Thing(1) in Local Adapter(1) Needs to Execute Action of Remote Thing(2) on Adapter(2)

- Thing(1) on Adapter(1) executes action of remote Thing(2) on Adapter(2) by calling Agent API service POST /remote/objects/vicinity-oid/actions/aid where
  - vicinity-oid is the VICINITY identifier of remote Thing(2)
  - aid is identifier of the action, provided in its Thing Description
  - Adapter(1) identifier and local identifier of Thing(1) are attached in request headers
- The request reaches the Gateway API on the remote VICINITY node
- The remote Gateway API executes the action and generates specific *task-id* for the current action execution. If there is another task running for this action, task is queued.

- The Gateway API passes each task execution request to the Agent, which performs translations and passes the execution request into Adapter(2).
- To execute the action, the Agent calls Adapter(2) API endpoint specified in Thing(2) Description to execute its action.
- Adapter(2) executes action on Thing(2). Execution is now running.
- *task-id* generated for this specific action execution is returned back to Adapter(1). Adapter(1) needs to remember this task-id and its relation to action being executed in order to get the result of the action task execution later.

Use-Case: Thing(2) in Adapter(2) Continuously Updates the Status of Task Execution

While the task is running, Thing(2) may continuously update the status of task execution, including current action output (e.g. for "fade-in" action in the Light Bulb example, current output can be actual *light intensity*).

- Adapter(1) checks the action execution status by calling Agent API service: GET /remote/objects/vicinity-oid/actions/aid/tasks/task-id, where:
  - *vicinity-oid* is the VICINITY identifier of Thing(2)
  - *aid* is identifier of action
  - *task-id* is identifier of current task execution within the action *aid*
  - Adapter(1) identifier and local identifier of Thing(1) are attached in request headers
- Response retrieved from the remote Gateway API contains the action output with the current execution status, which is one of the nominal values:
  - running: the action task with the identifier *task-id* is currently running
  - finished: the task was executed successfully
  - failed: the task execution failed
  - pending: another task is currently executed, the task is waiting for the running task to finish
- Adapter(1) is the one who is responsible for processing action status response.

9.3   Events

The VICINITY implements classic publish-subscribe eventing strategy. Things may publish data into event channels once the conditions are met. Subscribers automatically receive data from event channels, to which they are signed-in. Data are received automatically as they are published, without explicit request.

If Thing needs to publish event data, it first needs to open specific event channel, where data will be pushed. If Thing needs to automatically receive event data, first it needs to subscribe to specific channel, where data are pushed.

In the VICINITY, each event channel has its unique name. The name of event channel is always in the form: /objects/vicinity-oid/events/eid where:

- *vicinity-oid* is Thing VICINITY oid of publisher
- *eid* is identifier of the event provided in Thing Description

Combination of the Thing VICINITY oid and the event identifier ensures, that name of each event channel will be unique.

Management of event channels can be done either automatically or dynamically on demand. The Agent can be configured to automatically open or subscribe to the event channels. Adapter may the use Agent API to open or subscribe to event channel by request. All common scenarios will be described in the following.

Use-Case 1: Thing(1) in Local Adapter(1) Needs to Open Event Channel Automatically

Event channel can be automatically open by extending the Agent configuration for Adapter(1). The following configuration of *Access Point* will automatically open channel for event "power-consumption" for Thing(1) in Adapter(1).

```
{
  "credentials": {},
  "adapters": [
    {
      "adapter-id": "adapter-1",
      "endpoint": "http://adapter-1.endpoint",
      "events": {
        "channels": [
          {
            "infrastructure-id": "thing-1-local-oid",
            "eid": "power-consumption"
          }
        ]
      }
    }
  ]
}
```

Use-Case 2: Thing(1) in Local Adapter(1) Needs to Open Event Channel Dynamically

Event channels may be opened on demand, using the Agent API. To open event channel for event "power-consumption" of Thing(1) on Adapter(1) it is required to use the Agent API service: **POST /events/power-consumption**; Adapter(1) id and local Thing(1) identifiers are sent in request headers. The Agent performs automatic translations into common VICINITY identifiers *VICINITY oid* and opens the channel.

Use-Case 3: Automatic Subscription of Thing(1) of Local Adapter(1) to the Event Channel Opened by Thing(2) of Adapter(2)

Event channel can be automatically subscribed by extending the Agent configuration for Adapter(1). The following configuration of *Access Point* will automatically subscribe Thing(1) of Adapter(1) to channel for event "power-consumption" opened by remote Thing(2) of Adapter(2).

```
{
  "credentials": {},
  "adapters": [
    {
      "adapter-id": "adapter-1",
      "endpoint": "http://adapter-1.endpoint",
      "events": {
        "subscriptions": [
          {
            "infrastructure-id": "thing-1-local-oid",
            "oid": "thing-2-vicinity-oid",
            "eid": "power-consumption"
          }
        ]
      }
    }
  ]
}
```

Use-Case 4: Dynamic Subscription of Thing(1) of Local Adapter(1) to the Event Channel Opened by Thing(2) of Adapter(2)

Event channels may be subscribed on demand, using the Agent API. To subscribe Thing(1) on Adapter(1) to listen to "power-consumption" events published by remote Thing(2) in Adapter(2), it is required to use the Agent API service: **POST /objects/vicinity-oid/events/power-consumption** where "vicinity-oid" is the VICINITY identifier of Thing(2). Adapter(1) id and local Thing(1) identifiers are sent in the request headers. The Agent performs automatic translations into common VICINITY identifiers and subscribes Thing(1) to the channel.

Use-Case 5: Thing(1) of Local Adapter(1) Needs to Publish Event Data

To be able to publish data, channel for Thing(1) and corresponding event must be open. Thing(1) in Adapter(1) may publish data for event "power-consumption" by using the Agent API: **POST /events/power-consumption**; Adapter(1) id and local Thing(1) identifier are sent in request headers. The Agent performs automatic translations into common VICINITY identifiers and publishes data to corresponding channel opened for Thing(1).

Use-Case 6: Thing(1) of Local Adapter(1) Receives Event Data Published by Remote Thing(2) of Adapter(2)

To be able to consume data from channel for remote Thing(2) of Adapter(2), the channel must be open and Thing(1) must be subscribed to this channel. The Agent automatically passes the event data to event channels subscribed by Adapter(1). To receive event data, Adapter(1) must implement the following API service: **PUT: /objects/local-oid/publishers/vicinity-oid/events/eid** where:

- *local-oid*: local identifier of Thing(1) that is subscribed to event
- *vicinity-oid*: VICINITY identifier of remote Thing(2) that publishes the event
- *eid*: identifier of event

In our example, when the event is published, the Agent calls the following end-point on Adapter(1): **PUT: /objects/thing-1-local-oid/publishers/thing-2-vicinity-oid/events/power-consumption**. The Adapter is responsible to process data passed by the Agent via this API service.

10 Conclusions

This chapter provides a detailed and step-by-step integration guide to the VICINITY platform. Practical examples are given in order to demonstrate the usage of the VICINITY components for real scenarios. The reader learns how to interact with the Neighbourhood Manager, the Agent and the Gateway in order to register an Organisation with its devices and/or services and interact with other registered Things in the platform. Moreover, instructions are given regarding the implementation of the Adapter and the Thing Description which is the only user-specific software that an integrator has to implement in order to use the VICINITY Platform. The examples are also highlighting the semantic interoperability between different IoT Things and the range of privacy levels that are offered, which are the main unique selling points of the VICINITY Platform.

Index

© Springer Nature Switzerland AG 2021
C. Zivkovic et al. (eds.), *IoT Platforms, Use Cases, Privacy, and Business Models*,
https://doi.org/10.1007/978-3-030-45316-9

Printed in the United States
by Baker & Taylor Publisher Services